2002

Brownfield sites:
ground-related risks for buildings

JA Charles RC Chown KS Watts
BRE Centre for Ground Engineering and Remediation

G Fordyce
NHBC

constructing the future

Prices for all available
BRE publications can be
obtained from:
CRC Ltd
151 Rosebery Avenue
London EC1R 4GB
Tel: 020 7505 6622
Fax: 020 7505 6606
email:
crc@emap.co.uk

BR 447
ISBN 1 86081 571 5

First published 2002

Published by
Construction Research
Communications Ltd
by permission of
Building Research
Establishment Ltd

Acknowledgements

This report is the result of a research project carried out under DTI Partners in Innovation Contract cc1953 by:
Building Research Establishment Ltd (BRE)
with project partners:
National House Building Council (NHBC)
Construction Research Communications Ltd (CRC).

The BRE/NHBC project team was guided by a steering group, which comprised:

Mr David Baker	House Builders Federation
Mr David Beacham	Allott and Lomax
Mr John Cover	Bryant Homes
Mr Tim Everett	London Borough of Sutton
Mr Emyr Poole	English Partnerships
Mr Jonathan Steeds	W S Atkins
Mr Peter Witherington	ENSR International Ltd

A draft of the report was submitted to a Review Panel and, in addition to a helpful response from members of the Steering Group, valuable comments were received from:
Professor Stephan Jefferis University of Surrey.

Helpful comments on the final draft were made by:

Dr Paul Tedd	BRE
Mr R M C Driscoll	BRE

Contents

Glossary

Biodegradation
The decomposition of organic matter normally under anaerobic conditions.

Brownfield land
Land that has been previously developed for urban uses – see *Previously-developed land*

Collapse settlement
The settlement which occurs when a partially saturated soil undergoes a reduction in volume that is attributable to an increase in moisture content without there necessarily being any increase in applied stress.

Colliery spoil
Waste from the deep mining of coal.

Conceptual ground model
The conceptual ground model for a site has several elements, including a three-dimensional stratigraphic model of the ground, an understanding of the history of the site and of the ways in which it has been affected by various types of human activity, a groundwater model, and a soil contamination model.

Contaminated land
Part IIA of the Environmental Protection Act 1990: '*Contaminated land is identified on the basis of risk assessment. In accordance with the provisions of Part IIA and statutory guidance, land is only contaminated where it appears to the authority, by reason of substances in, on or under the land, that: (a) significant harm is being caused or there is a significant possibility of such harm being caused; or (b) pollution of controlled waters is being, or is likely to be, caused*'. Harm is defined by reference to harm to health of living organisms or other interference with the ecological systems of which they form a part and, in the case of man, is stated to include harm to property.

Desk study
An examination of existing information concerning a site, including historical records, geological maps, borehole records, and air photographs to determine ground conditions and previous land use.

Dynamic compaction
A ground treatment method in which deep compaction is effected by repeatedly dropping a heavy weight onto the ground surface from a great height.

Hazard
A situation or event that has the potential for harm, including human injury, damage to property, environmental and economic loss.

Low-rise buildings
Buildings not more than three storeys in height.

Mitigation
The limitation of the undesirable consequences of a situation or event.

Opencast mining
Mining carried out by excavation from the ground surface.

Previously developed land
In PPG3 (Department of the Environment, Transport and the Regions, 2000a), previously-developed land is defined as land that is or was occupied by a permanent structure (excluding agricultural or forestry buildings), and associated fixed surface infrastructure. The definition covers the curtilage of the development and includes land used for mineral extraction and waste disposal where provision for restoration has not been made through development control procedures.

Risk
A combination of the probability or frequency of the occurrence of a defined hazard and some measure of the magnitude of the consequences of the occurrence. The risk to a specified receptor can be defined as the product of the hazard and the exposure.

Risk management
The overall application of policies, processes and practices dealing with risk.

Risk register
The file where risk information is stored which includes a description of the risk, an assessment of its likelihood and consequence, and any remedial actions.

Vibrated stone columns
Deep vibratory ground treatment achieved by penetration into the ground of a large vibrating poker; usually the cylindrical hole formed by the vibrator is backfilled in stages with stone, forming stone columns. Often called '*vibro*'.

Walk-over survey
At an early stage a thorough visual survey of a site carried out to obtain information on ground conditions and land use.

Abbreviations

AGS	Association of Geotechnical and Geoenvironmental Specialists
BRE	Building Research Establishment
BS	British Standard
BSI	British Standards Institution
CEN	European Committee for Standardisation
CIRIA	Construction Industry Research and Information Association
DEFRA	Department of the Environment, Food and Rural Affairs
DTI	Department of Trade and Industry
DPC	Damp-proof course
EA	Environment Agency
EN	European standard
ENV	European pre-standard
HDPE	High density polyethylene
HSE	Health and Safety Executive
ICE	Institution of Civil Engineers
ICRCL	Interdepartmental Committee on the Redevelopment of Contaminated Land
LCR	Standard Land Condition Record
MDPE	Medium density polyethylene
NHBC	National House Building Council
pfa	pulverised fuel ash
SNIFFER	Scotland and Northern Ireland Forum for Environmental Research
UK	United Kingdom
USEPA	United States Environmental Protection Agency
VOC	Volatile organic compound
WRC	Water Research Centre

Part 1 Introduction

The increasing demand for new homes and the expansion of the commercial sector have to be met on a diminishing landbank, and the use of previously developed land for building development offers substantial advantages in social, economic and environmental terms. Furthermore, the sustainability agenda requires the long-term productive use of previously used land, which is often described as 'brownfield'. Following a description of the background to the report, its objectives, scope and structure are outlined. The report is concerned with those aspects of risk management for building developments on brownfield sites that involve ground-related hazards for the built environment. The report is, of necessity, concerned with hazards and risks, but it is emphasised that an unwarranted over-sensitivity to risk will, of course, defeat the objective of locating building developments on brownfield sites. The hazards to the human population from contaminated land are not discussed; guidance for the safe development of housing on land affected by contamination can be found in the Environment Agency and NHBC (2000) report.

1.1 Background to this report

To promote regeneration and minimise the amount of greenfield land being taken for development, Government is committed to maximising the re-use of previously developed land. A national target has been set: by 2008, 60% of additional housing should be on previously developed land (Box 1) and through conversions of existing buildings (Department of the Environment, Transport and the Regions, 2000a). This target has to be met against a background of social changes which could mean that up to 3.8 million extra households have to be accommodated by 2021 (Department of the Environment, Transport and the Regions, 2000b). The Urban Task Force (1999) considered that achieving the 60% target was fundamental to the health of society and recommended that ambitious targets should be set for the proportion of new housing to be developed on recycled land in urban areas where housing demand is currently low.

Building on previously developed sites (often termed *'brownfield'*) is in the public interest but there are risks associated with previous land use. Ground-related hazards for the built environment include poor load-carrying properties of the ground, interaction between building materials and aggressive ground conditions, gas generation from biodegradation of organic matter within the ground, and combustion. If such hazards are not identified, or are wrongly diagnosed, remedial measures may be required with significant implications for whole-life costs. The situation becomes more difficult and complex where there are also hazards for human health and detrimental effects on the natural environment and resources such as groundwater.

Box 1 Definition of previously developed land

In Planning Policy Guidance Note PPG3 (Department of the Environment, Transport and the Regions, 2000a), previously developed land is defined as land that is or was occupied by a permanent structure (excluding agricultural or forestry buildings), and associated fixed surface infrastructure. The definition covers the curtilage of the development and includes land used for mineral extraction and waste disposal where provision for restoration has not been made through development control procedures.

Using the wide definition in Box 2, brownfield land can be found in many different forms. Evidence of contamination is clear in the site in Figure 1. In contrast, the development in Figure 2 is on a restored opencast mining site which is unlikely to be contaminated; however, previous use of the land may have left a substantial depth of fill in a loose condition with the potential to cause major settlement of the ground surface. In the case shown in Figure 2, the ground was pre-loaded with a surcharge of fill prior to development.

Although brownfield land is a world-wide phenomenon, the issues associated with the redevelopment of such sites are particularly acute for Great Britain, a heavily populated island with a long industrial history. The new National Land Use Database has indicated that, in England, some 58,000 ha of brownfield land is either vacant, derelict or available for redevelopment (Department of the Environment, Transport and the Regions, 2000b).

The scale of the problem is further illustrated by the size of the £1 billion plus package announced in 1996 for the regeneration of major coalfields; it deals only with brownfield sites associated with coal mining. It was announced that

Box 2 Definition of brownfield

The term '*brownfield*' has been widely used, but it has no universally recognised definition. Alker et al (2000) have summarised many of the definitions that have been proposed. Two different concepts should be noted.

- The United States Environmental Protection Agency (USEPA) has defined brownfields as abandoned, idled or under-used industrial and commercial facilities where expansion or redevelopment is complicated by real or perceived environmental contamination.
- In the UK, brownfield commonly has been understood as signifying the opposite of '*greenfield*' in planning terms, where greenfield is taken to mean land that has not been previously developed.

The USEPA definition would equate brownfield with contaminated sites and this seems unduly restrictive. The contrast with greenfield provides a more appropriate starting point for a definition of brownfield in the United Kingdom.

The Parliamentary Office of Science and Technology (1998) commented that because there is no agreed definition, many practitioners regard brownfield land as any land that has been previously developed, and it is in this sense that the term is used in this report. It should be noted that 'brownfield' represents a broader concept than derelict land.

some 910 ha of land were to be reclaimed for residential, commercial and retail uses (Department of the Environment, 1996; Sleep, 1996). Many of these sites will involve building on colliery spoil (Skinner et al, 1997).

The scale of proposed future development on brownfield land means that extra costs associated with ground-related problems could be substantial. Such costs present potentially damaging obstacles to the furtherance of the policy of encouraging the location of building developments on brownfield sites. The risks for the planning authorities, developers and insurers provide a significant deterrent to developing these sites and the long-term risks for occupiers and owners also need to be addressed.

It has been questioned whether current approaches to the sustainable redevelopment of brownfield land are always suitable. In some cases, hazards may be overlooked or their significance is not recognised but, in other cases, solutions may be over-engineered. Many of these issues have been debated by Wood and Griffiths (1994). The way that risks are perceived can be of crucial importance and this is considered in Part 3.2.

There can be many hazards on brownfield sites and it is vital at an early stage to identify the most significant problems and to evaluate the risks that they pose. It is necessary to define the tolerable level of risk. On housing developments, risks to human health from contamination may be a significant issue but this should not distract attention from the hazards to the built environment. Although there is a need for improved risk management, it should be emphasised that the redevelopment of brownfield sites can have significant benefits and that greenfield sites are not necessarily problem free.

Where building development is proposed, risk assessment should consider a broad spectrum of potential receptors which may be vulnerable to different hazards. Three systems may be at risk in brownfield developments:

- the human population (Box 3);
- the natural environment (Box 4);
- the built environment (Box 5).

The three systems described in Boxes 3, 4 and 5 are, of course,

Box 3 System at risk – human population

In an increasingly risk averse society, matters concerning health and safety receive much attention and publicity. As a consequence, and since brownfield sites are sometimes contaminated, hazards associated with contamination and the risks posed to human health have been the dominant issues in the redevelopment of derelict land and brownfield sites. This preoccupation was illustrated in a paper on risk assessment and soil contamination in which it was affirmed that: *Risk assessment is appraising the possibility and severity of potential adverse health events* (Aldrich et al, 1998). No mention was made of the potentially adverse impacts of contamination on the various aspects of the natural environment or the built environment.

From the early stages of investigation through to the final use of the site, a range of people may be at risk. Where low-rise housing is built on the land, the occupiers will be the people most at risk from many of the hazards. While there is no evidence to suggest that chemical contamination on brownfield sites has posed a major threat to human life, at least in the short term, long term health and quality of life could be affected and this is difficult to evaluate.

Box 4 System at risk – natural environment

Threats to the natural environment are often narrowly conceived to concern soil and ground-water contamination. However, a much wider range of issues may need to be examined in the light of growing concern over degradation of the natural environment and the increasing prominence of environmental pressure groups promoting concepts such as biodiversity. Proposals to reclaim derelict land may lead to serious difficulties with opposition from those who consider that eco-systems established amidst the dereliction, and present as a direct result of it, should be preserved. For such sites, it is difficult to establish an appropriate balance between economic well-being and the natural environment because, unlike matters of human health and building damage, there are no widely agreed objectives or ground rules.

Box 5 System at risk – built environment

The site and the building development form an interactive system and it is important to evaluate the risk of adverse interactions during the lifetime of the development. Hazards to the built environment on a brownfield site can be physical, chemical or biological in character and concerns can include the following:

- poor load carrying properties of the ground
- interaction of building materials and services with aggressive ground conditions
- gas generation from biodegradation of organic matter and from other deleterious substances in the ground
- combustion.

These hazards are described in Part 2.

interdependent. While the degradation of the natural environment and sustainable construction involving the rate of use of renewable and non-renewable resources, and the emission of pollutants are important issues, the primary objective is to build safe, durable and economic structures.

Figure 1 Contaminated land

Figure 2 Low-rise housing built on opencast mining backfill

1.2 Objectives of this report

The primary objective of this report is to assist the planning and design processes for building developments on brownfield sites by presenting planners, building professionals and insurers with guidance on those circumstances and processes that represent a hazard to the built environment and a methodology for taking risk-based decisions on available options for development. Generally, brownfield sites present a greater risk to the built environment than greenfield sites because many of the hazards found on greenfield sites have been extensively researched and are better understood; also they are more easily identified and are more familiar to builders.

A secondary objective of the work on which this report is based was to form a database of incidents and problems affecting buildings on brownfield sites. This should facilitate the assessment of the likely incidence and scale of hazards where problems arise some years after building development has been completed and should enable future research to be more realistically focused on areas that have been shown to be problematical. A representative group of seven case histories from the database has been included in Part 4.

1.3 Scope of this report

Although it is recognised that human health and other issues regarding the biosphere are of great concern, it is the ground-related risks to the built environment which are the subject of the guidance presented here. The built environment includes:

- buildings
- transport infrastructure;
- services;
- gardens and ancillary buildings.

The report outlines a systematic means of managing the range of risks to the built environment over the lifetime of the building development. Decisions which may be required include, for example:

- for foundations on a deep fill, whether to adopt a low-cost ground improvement technique in preference to piling, taking into account possible remedial costs within the design life of the building;
- for foundations where there is a gas migration problem, whether a passive venting system is adequate;
- for services in contaminated ground, the level of protection required.

The guidance in this report is principally focused on low-rise buildings, and particularly housing. It is seen as complementary to, firstly, the model procedures for the management of contaminated land being developed by DEFRA/Environment Agency and, secondly, the Environment Agency and NHBC (2000) guidance for the safe development of housing on land affected by contamination. These procedures and guidance are briefly described in Part 3.1; reference is made to the guidance given on hazards to the human population from contaminated land but such guidance is not repeated here.

Guidance is concerned with the technical aspects of building development on brownfield sites rather than with legal issues.

1.4 Structure of this report

● The major ground-related problems are described in Part 2 where four basic types of hazard for the built environment are identified with brief descriptions. There are references to more detailed accounts of these hazards.

● The risk management process is outlined in Part 3. This makes use of the information on ground-related problems presented in Part 2 and also the case histories in Part 4. The principal elements of the risk management process are identified and described. Two key components of the process are the development of a conceptual model of the ground conditions and the compilation of a risk register.

● Seven case histories are presented in Part 4 to illustrate commonly occurring problems and their impact on building development and to identify the deficiencies in the approach to risk which caused or exacerbated the problems.

● A list of references is in Part 5, on page 42.

Part 2 Ground-related problems

All buildings and other constructed facilities come into contact with the ground; unforeseen ground-related problems often lead to cost increases and delays during construction. Furthermore, ground-related problems may emerge many years after the completion of construction. The principal types of ground-related hazards on brownfield sites are identified as ground movement, vulnerability of construction materials to aggressive soil conditions, gas migration and subterranean fires. Brief descriptions are given of each of these hazards and references provided to more detailed accounts.

2.1 Principal ground-related hazards

For many years, ground problems have been attributed to inadequate site investigation and, doubtless, there have been many cases where site investigation has not been undertaken with due diligence. However, there are also many cases where substantial site investigations have been undertaken, and yet unexpected ground-related problems have still occurred.

The importance of an appropriate site investigation, including desk study, walk-over survey and intrusive ground investigation, in identifying the principal hazards cannot be over-emphasised, but the uncertainty of ground conditions means that risk management is essential (Clayton, 2001a and 2001b). This has increasing significance in many parts of the country where most new housing developments take place on land where previous use has left a wide range of physical and chemical hazards. In the past, when most building development was on greenfield sites, geological maps provided some indication of the types of problem that might be encountered but, as development becomes predominantly on brownfield land, this is no longer the case. Site investigation is described in Parts 3.3 and 3.4, and it is emphasised that a clearly focused site investigation is required that develops out of detailed desk studies (Association of Geotechnical and Geoenvironmental Specialists, 1998a).

There can be many different hazards on brownfield sites and many of them will be related to the previous use of the site; there may also be problems related to the original state of the ground. These can include various types of ground movement and aggressive ground conditions. Problems such as those associated with rising ground water levels due to reduced abstraction by

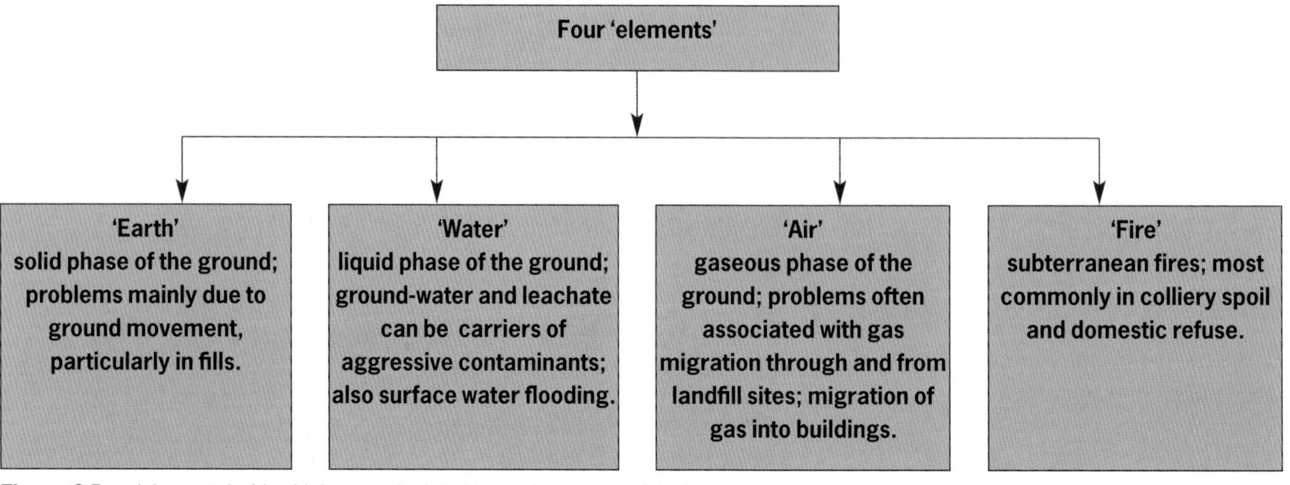

Figure 3 Four 'elements' with which ground-related hazards are associated

declining manufacturing industries, and with flooding hazards on floodplain developments, can affect both greenfield and brownfield sites.

It may be helpful to consider the ground-related hazards as being associated with the four 'elements' of the ancients, namely earth, water, air and fire – Figure 3. In this scheme, the first three 'elements', earth, water and air, can be thought of as representing the solid, liquid and gaseous phases of the ground and the fourth 'element', fire, as representing subterranean fires.

There is much literature on the problems encountered on brownfield sites, including proceedings of many conferences, some with a geotechnical focus (for example, Yong and Thomas, 1997 & 2001; Seco e Pinto, 1998). Many of the problems have been caused by excessive ground movements; physical problems include buried foundations and settlement of filled ground. There is a vast range of problems associated with chemical contamination which can present an immediate or long-term threat to human health, plants, amenity, construction operations and buildings and services. Biodegradation of organic matter may lead to the generation of gas.

Some of the hazards that can be encountered on brownfield sites are shown in Figure 4. In view of their diversity, it is helpful to adopt some form of classification and the major hazards to the built environment are here reviewed under the following four headings:

- ground movement;
- vulnerability of construction materials to aggressive ground conditions;
- gas migration;
- subterranean fires.

As with all classifications, this has its weaknesses but it facilitates an orderly presentation of the hazards. Information on different aspects of these hazards can be found in the Parts of this report listed in Table 1.

Table 1 Parts of this report dealing with different hazards

Hazard	Description	Identification	Treatment	Foundations and services design
Ground movement	2.2	3.4A	3.7A	3.8A
Vulnerability of construction materials	2.3	3.4B	3.7B	3.8B
Gas migration	2.4	3.4C	3.7C	3.8C
Subterranean fires	2.5	3.4D	3.7D	3.8D

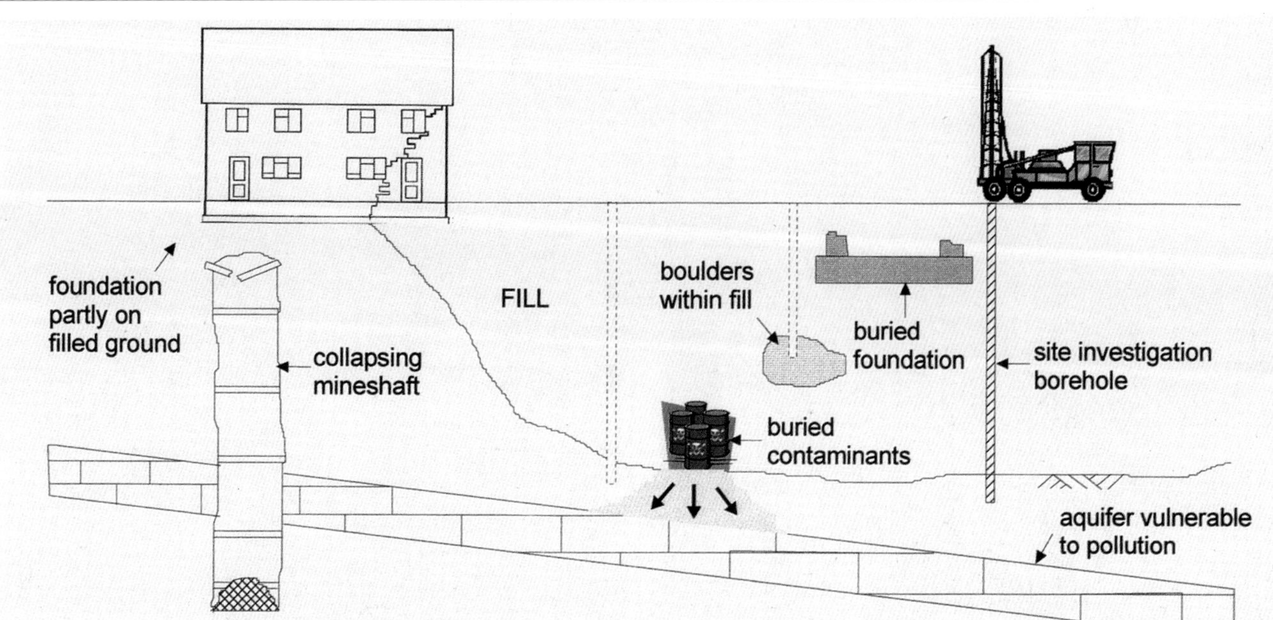

Figure 4 Typical hazards on brownfield sites

2.2 Ground movement

The foundations and superstructure of buildings can be damaged by excessive ground movements. Settlement (downward movement) is the most common form but, in certain situations, the ground may heave. Horizontal ground movements can be particularly damaging. The term 'subsidence' is also used to describe downward movement of the ground supporting a building.

The load carrying characteristics of the ground on a brownfield site may be inadequate owing to a variety of causes, classified into two general types:

- The ground may have had poor load carrying properties before it was affected by human activity and the full range of ground hazards which may be found on greenfield land may also be found on brownfield sites. Problems in this category, which include building on highly compressible soft clay and organic soils, are covered in standard soil mechanics and foundation design text books (eg Budhu, 2000; Lambe and Whitman, 1979; Tomlinson, 1995) and are not discussed further here.

- Previous human activity on the site may have created ground-related problems. These are commonly, although not exclusively, related to the uncontrolled placing of fill material. Other hazards in the ground include buried foundations, old pipework and tanks from previous site usage, and old mine workings. It is this category of ground movement, which is related to human activity and peculiar to brownfield sites, that is described here.

Commonly occurring problems are related to fill being deeper and more variable than anticipated and the failure to accurately locate the limits of backfilled areas, such as infilled quarries and pits. The problems that this can cause are shown by case histories 3 and 4 in Part 4. Figure 5 shows a house built on deep fill seriously damaged by ground movements.

Figure 5 House seriously damaged by ground movements

Commonly encountered types of ground movement on brownfield sites are described under the following headings: compressible fills, expansive fills, biodegradable fills, buried foundations and infrastructure, and shallow mine workings. This is not an exhaustive list and particular types of previous land use may cause other forms of ground movement. For example, in the Peterborough area where brick kilns have been located on the Oxford Clay, the heat of the kilns has dried the clay to substantial depths (Cooling and Ward, 1948). When the kilns are removed, ground heave can be expected over a long period.

Compressible fills

Little control may have been exercised over the placing of fill material and the poor load carrying properties of many non-engineered fills have been associated with their heterogeneity and their loose, poorly compacted condition – Digest 427 and BRE Report BR 424. The major geotechnical problem is usually associated with settlement of the fill due to effects other than the weight of the building.

While some time-related creep settlement will occur in most fills, the major concern is usually related to poorly compacted or excessively dry fill which is likely to be vulnerable to a reduction in volume when the moisture content of the fill is increased. This phenomenon, termed 'collapse settlement' or 'collapse compression', on wetting can occur without any increase in applied stress (Charles and Watts, 1996). It can occur at depth within the fill or close to ground level, due either to a rising groundwater table or to downward infiltration of surface water. As it is differential settlement rather than total settlement that damages buildings, local collapse compression from a surface source of water is of particular concern. It will not usually be feasible to prevent some infiltration of water from surface sources, such as leaking drains and water mains, over the lifetime of a building. In low lying land, inundation could occur due to flooding.

Settlement may be related to the depth of fill and, therefore, differences in the depth of the fill which occur under a building are of significance. Differential settlement is likely to be particularly marked at the edge of a backfilled area and founding buildings partly on fill and partly on natural ground should be avoided (case history 3). Guidance on the size and location of the building exclusion zone adjacent to the edge of a backfilled opencast mine has been given by Charles and Skinner (2001).

Reduction in volume is usually due to volumetric compression of the fill but there are cases where fills reduce in volume owing to loss of material. This loss can be caused by:

- erosion of fine particles;
- dissolution of soluble chemical waste.

Serious foundation problems caused by soluble chemical wastes have been described by Greenwood (1986) and Vadgama (1986).

Expansive fills

A number of ground conditions encountered on brownfield sites have the potential for ground movements induced by chemical reactions, usually expansion. One of the most common is the presence of slags from iron and steel-making processes. Slags may be present as a result of disposal as waste or use as fill or hardcore.

Two main types of slag are produced. Blastfurnace slag arises when iron ore is smelted to produce pig iron, and steel slag arises when pig iron is converted into steel. Slags are also produced during the production of special steels and alloys but the quantities are small. The principal hazard from the presence of slags is that they may expand, possibly decades after deposition, causing damage to buildings, roads or other structures; not all slags are expansive. There are other hazards associated with slags, which may contain small quantities of toxic elements, but these are not dealt with here.

Steel slags may contain phases that cause expansion on wetting (eg free lime, free magnesia). The expansive properties of steel slags, including those that have been stockpiled for many years, are well known. There have been numerous failures of structures built on such slags, in the United Kingdom and overseas. However, as these failures have often led to legal actions, little information about the particular circumstances is in the public domain. Warnings about the potential problems were first published by BRE in 1981 and are included in the current edition of Digest 276. Further information on the problem of expansive slags is given in the Environment Agency Report P331 (Garvin et al, 1999).

Long-term exposure to the elements does not guarantee that steel slags do not retain a potential to expand. Slag that has spent several decades in stockpiles may expand in later years if disturbed. Possible enhancing factors include:

- ingress of moisture to previously dry materials; there is sometimes a cemented surface layer which has inhibited water ingress;
- ingress of air permitting the oxidation of sulfide to sulfate;
- bringing into contact slags of different composition or stages of degradation that chemically interact with one another.

Biodegradable fills

Where fills contain significant quantities of putrescible material, large reductions in volume can occur. The composition of household refuse has changed markedly over the years. Before 1960, the inert ash content was over 50%; between 1960 and 1970 this reduced to about 20% with subsequent further reduction. The settlement of old and recent municipal refuse has been measured at a number of sites by BRE (Watts and Charles, 1999).

The volume reduction on biodegradation is accompanied by gas generation and the formation of leachate. Leachate can pollute aquifers and cause other environmental problems. The hazard for the built environment posed by gas migration may preclude housing developments and this subject is examined in Part 2.4. Problems can arise with building developments on small uncontrolled tips where the presence of wood or other biodegradable material has not been identified.

Buried foundations and infrastructure

Buried within a brownfield site there may be heavy foundations, pipework and storage tanks remaining from previous use of the site. These may form obstructions when excavating for foundations for the new development (Figure 6) or during piling (case history 2). If left in place, they can form local 'hard spots' which cause differential settlement.

It is often inconvenient and expensive to avoid obstructions by siting buildings away from them, or to remove or bridge over old foundations. On redevelopment sites with severe space and financial constraints, re-use of existing foundations could have substantial benefits to clients and further the progress of sustainable construction (Part 3.8).

Shallow mine-workings

In areas where there is a history of mining, the presence of large voids at shallow depths should be considered. Detailed information on building in mining localities is in CIRIA Special Publication 32 (Healy and Head, 1984) and some more general advice is in Atkinson (1993). Building houses on land which is underlain by known shallow coal workings or other mineral workings can result in substantial development costs. Case history 6 describes an incident where this particular hazard was overlooked at the site investigation stage.

2.3 Vulnerability of materials to aggressive ground conditions

Some form of contamination may be present at brownfield sites and aggressive ground conditions may be associated with this. Aggressive ground conditions may exist naturally but at brownfield sites building materials are likely to be subjected to aggressive environments which make them liable to physical or chemical changes: natural aggressive conditions may be exacerbated by the contamination. Such changes may result in loss of strength or other properties that put at risk their structural integrity or ability to perform to design requirements.

Figure 6 Underground obstructions and buried foundations

Aggressive environments include severe climates, coastal locations and polluted atmospheres as well as aggressive soils. They can result in increased maintenance requirements, a reduction in the service life of materials and buildings and ultimately risks to health and safety or the environment. There can be damage to materials incorporated in the building to protect people or the environment; for example, plastics membranes used to protect against gas.

The interaction between construction materials and chemically aggressive ground has been studied at BRE; Report BR 255 gives general guidance on the performance of building materials used in contaminated land. Materials at risk include concrete, mortars, metals, plastics and masonry. The Environment Agency has produced a procedure for the assessment and management of risks to buildings, building materials and services from contaminated land (Garvin et al, 1999). The requirements for concrete buried in the ground to resist sulfate and acid attack were given in Digest 363 which has now been updated as Special Digest 1.

The following factors influence whether deterioration will take place and, if it does, the rate of deterioration.

- The use of unsuitable materials, for example the wrong choice of cement or aggregates for mortar and concrete.
- Poor production or placing of concrete that leaves the concrete with a more open, permeable texture, lacking in cover to reinforcement.
- Placing a vulnerable material in an aggressive environment without appropriate protection from ground conditions.

In aggressive soils the potential for contaminant attack depends upon:

- The presence of water as a carrier of aggressive contaminants (except in the case of free-phase organic contamination). Soils do not have to be saturated: lower moisture contents can reduce the amount of water available for contaminant migration but may increase the concentration of dissolved species.
- The availability of the contaminant in terms of concentration, solubility and replenishment rate of the aggressive solution.
- Contact between the contaminant and the building material.
- The sensitivity of the material to the contaminant, (the inherent durability of the material and the properties that cause it to react with the contaminant).

Structural elements

The effects of sulfates, acids and chlorides give rise to concern for both unreinforced and reinforced concrete foundations or other exposed elements (for example, material below the dpc or exposed to the sea). Magnesium, ammonia and phenol are also known to cause deterioration of concrete. Sulfate attack on concrete is characterised by expansion, leading to loss of strength and stiffness, cracking, spalling and eventual disintegration. High quality dense concrete is a prerequisite for durability and there is no substitute for the quality of the materials as placed; poor compaction is possibly the major problem together with excessive water/cement ratio. The durability of buried concrete is also a function of cement content; high cement content concrete with low water/cement ratio is the most durable.

Damage to materials can lead to weakening of foundation concrete and could compromise the stability of the building but the consequences of chemical attack on concrete foundations are not easy to assess. Localised deterioration and cracking will result in the need for remedial work.

Services

Buried services are particularly vulnerable to damage and degradation. These include concrete, metal, plastics and ceramic pipes.

- A damaged water supply pipe might allow the entry of contaminants into the water supply.
- A damaged gas pipe could release gas into the surrounding ground.
- An undamaged plastics pipe may be permeable to organic contaminants and gases.

The potable water supply is of particular concern; case history 7 describes an example of this type of problem. Plastics pipe performance in contaminated land has been the subject of a WRC report (Stephens and Norris, 1994). Organic chemicals are more likely to attack plastics pipes: metal pipes are preferred in these conditions. The permeation of plastics pipes by organic solvents and hydrocarbons can be a significant problem. Polyethylene pipes with an aluminium foil layer have been developed to stop the ingress of hydrocarbons. The pipes comply with the Water Supply (Water Fittings) Regulations (Stationery Office, 1999). Fuel oils are the major source of contamination of liquids in plastics pipes. Inorganic contaminants are more likely to attack metal or cement-based pipes. No single pipeline material is immune from all contaminants. Guidance on the selection of pipe materials has been given by Trew et al (1995).

2.4 Gas migration

Gases emanating from the ground, such as methane, carbon dioxide, radon and VOCs, can present hazards (see BRE Information Paper 2/87), including:
- asphyxiation;
- poisoning;
- explosion.

Where fills or natural soils contain biodegradable material, gas generation forms a significant hazard for building developments – Figure 7. Landfill sites can present a particularly serious hazard because wherever biodegradable material is deposited, microbial activity will generate landfill gas: a mixture of flammable and asphyxiating gases. The main constituents of landfill gas are methane and carbon dioxide:
- methane is an asphyxiant, it will burn and can explode in air;
- carbon dioxide is non-flammable, toxic and asphyxiant;
- many of the other components of landfill gas are flammable, all are asphyxiant, and some are toxic.

CIRIA report 152 has information on significant gas concentrations (O'Riordan and Milloy, 1995) as does Approved Document C of the Building Regulations (Department of Transport, Environment and the Regions, 2000d). The general approach to building development close to a landfill site is described in the Approved Document and this has been summarised in Part 3.8. Reference is made to guidance on the construction of buildings near but not on landfill sites published in BRE Reports BR 212 and BR 414 .

An assessment of the risk of lateral migration of gas into the ground surrounding the gassing landfill, requires an understanding of the geology of the area. An explosion which completely destroyed a bungalow and badly injured three occupants at Loscoe in Derbyshire on 24 March 1986 is believed to have been caused by the lateral migration of methane gas from a landfill site 70m from the bungalow (Williams and Aitkenhead, 1991).

Gas migration can be a hazard in other types of situation. For example, problems can arise as a result of the migration of the vapour phase following petroleum and solvent spillages or from leaking supply pipes. Gas can be associated with mining.

The worst disaster in the United Kingdom involving gas migration into a building occurred on 23 May 1984, when sixteen people were killed by an explosion in a water supply valve house at Abbeystead in Lancashire (Health and Safety Executive, 1985).

Ingress routes
1 Through cracks in solid floors
2 Through construction joints
3 Through cracks in walls below ground level
4 Through gaps in suspended floors
5 Through cracks in walls
6 Through gaps around service pipes
7 Through cavities in walls

Locations for gas accumulation
A Wall cavities and roof voids
B Beneath suspended floors
C Within voids caused by settlement of the ground
D Drains and soakaways

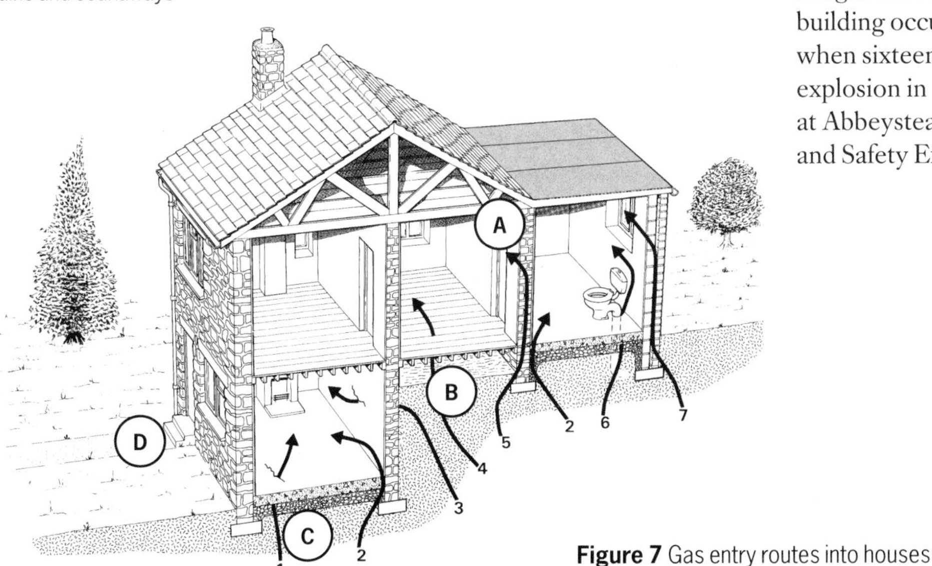

Figure 7 Gas entry routes into houses

2.5 Subterranean fires

There is little published information on the hazards arising from subterranean fires in brownfield sites and the effects of such fires on building materials and services. Garvin et al (1999) have summarised earlier work. A fire which occurred beneath a small estate of bungalows is reported to have caused subsidence and structural collapse. Surveys of the incidence of subterranean fires in the UK undertaken in the 1980s showed that the majority of fires occurred in sites containing domestic refuse and colliery spoil; the latter could be more frequently encountered as brownfield sites in mining areas are redeveloped. Geen (1944) described a fire on an extensive scale below a factory built on an old ironworks tip.

Hazards posed by subterranean fires to buildings include (Andrews, 1944 and BRE Information Papers 2/87 and 3/89):

- Settlement of the buildings and services. A smouldering fire occurring in combustible materials under a building will consume these materials and produce cavities. This may compromise the stability of the building to cause subsidence and deformation and fracture of service pipes.
- Heat damage to underground structures such as foundations and service pipes. The passage of a fire beneath a building may expose the construction materials and services to high temperatures. Temperatures of up to 1000°C can occur, which could cause concrete to spall and melt plastics service pipes.
- Acid gases causing a degradation of building materials and pipework. Depending on the nature of the combustible materials, a smouldering combustion can produce gases such as carbon dioxide, carbon monoxide, sulfur dioxide, hydrogen sulfide, and hydrogen chloride. If these attack the building materials over a long period, the integrity of the materials could be affected; some subterranean fires have been burning for more than 20 years.

Two situations need to be addressed:

- The hazard posed by an existing fire. While it is obviously unwise to build on a site where combustion is already taking place, there can be a hazard even where the seat of the fire is not directly under the buildings. A fire which starts at a location distant from the building could still pose a threat due to lateral migration through the combustible material of the site if there are no barriers to halt the passage of the combustion front.
- The risk that a subterranean fire will occur subsequent to building. Some ground conditions, including certain types of fill such as colliery spoil, may be susceptible to combustion. There is strong evidence that where colliery spoil is placed as an engineered fill and compacted so that air voids are reduced to a minimum, the risk of combustion is very low.

Part 3 Risk management

Ground-related risks for the built environment are substantial and not all risk for subsurface conditions can be avoided or eliminated. Ground conditions may be poor at brownfield sites and the risks for those involved in the development can be great. The risks are manageable and controllable provided that they are identified in time; an adequate understanding of brownfield site risks is essential. In many cases risks can be substantially reduced without major expenditure. For example, the problems of differential settlement can be mitigated by applying some form of ground treatment and/or providing stiff raft foundations for relatively little extra cost. A risk-based approach to ground-related hazards for the built environment is outlined based on the following steps; hazard identification, risk assessment, development of risk mitigation strategy including ground treatment, and foundation and services design. Two key components of the approach are the development of a conceptual model of the ground conditions and the compilation of a risk register.

3.1 Risk-based approach to ground-related hazards

Building and construction always involve uncertainties: nowhere are those uncertainties greater than in the ground (Clayton, 2001a and 2001b). An adequate understanding of brownfield site risks is essential, including a knowledge of the hazards described in Part 2, some assessment of the probability of their occurrence and an understanding of their consequences.

Ground conditions are often poor on brownfield sites, but the risks are manageable and controllable provided that they are identified in time (Hatem, 1998). Risk management for building projects is most effective when it is started during pre-project planning before the purchase of land for development. The client is likely to be most involved at this early stage of the development when project objectives are defined, a conceptual design is formulated and financial planning is carried out.

One unfortunate side effect of evaluating risk is that there is, of necessity, a concentration on problems and potentially negative effects. The re-development of brownfield sites can have many advantages and, while it is important to identify hazards and to evaluate risks, the benefits should not be overlooked. An over-conservative approach to risk and the perceived need to minimise future potential liabilities can lead to a loss of wealth to individuals and to the consumption of an unnecessarily large proportion of national resources. The key engineering issue is management of the risk at acceptable cost, not elimination of the risk whatever the cost.

With a continuing high proportion of building developments on brownfield sites with the various hazards that these may contain, risk assessment and risk management have major roles to play, particularly in determining the required level of remedial action to achieve a 'suitable for use' condition. Quantitative risk assessment has been widely applied in the chemical and nuclear industries, but has received less use in contaminated land. The source-pathway-receptor approach has been extensively used for assessing risk to the human population and this can be quantitative.

Quality is another important concept in brownfield development. Contaminated Land Research Report CLR 12, describes *A quality approach for contaminated land consultancy* (Environmental Industries Commission, 1997).

The successful outcome of a project requires not only technical competence, but also an appropriate organisational structure in which the responsibilities of the various parties are identified and which facilitates effective communication between those parties. The parties involved will depend on the form of the contract as well as the type and size of the project, but will normally include the developer or promoter of the scheme, the project

Box 6 The standard land condition record

The standard land condition record (LCR) is a voluntary log-book for brownfield land, kept and maintained by the landowner and transferred with the land. It uses standard forms for the collection and presentation of data in a technically robust and systematic way. Its primary role is to provide a clear record of the physical and chemical nature of the land and provide details of the use that the land has supported. Anyone with a vested interest in the site can view basic technical data, enabling them to make better informed decisions based on existing information. It should supplement the desk study, which is a key element in the planning and investigation phases of the redevelopment process. It will help focus the attention of developers and planners on the probable hazards of the site at the earliest possible stage of redevelopment, improving consistency and confidence in the information held and assisting negotiations with regulatory authorities, landowners etc.

To encourage high standards and improve market confidence in the redevelopment of urban land, the Institute of Environmental Management and Assessment manages an accreditation scheme for engineers and other professionals who independently check the technical data held in each LCR. Those on the register will be permitted to call themselves a specialist in land condition and will be the only people who have been independently recognised as being competent to sign off a land condition record (Bennett, 2002). This is an essential step in assuring the quality, consistency and validity of the LCR. Although independent verification means additional expenditure, the process of preparing and having an LCR approved should be no more expensive than preparing and carrying out a standard site investigation.

Box 7 Model procedures for the management of contaminated land

The legal and policy framework relies on a risk-based approach to identifying and managing contaminated land. The key components of the primary procedures of the risk-based approach are:

- Risk assessment Phase 1: hazard identification and hazard assessment.
- Risk assessment Phase 2: risk estimation and risk evaluation.
- Evaluation and selection of remedial measures.
- Implementation of risk management action.

designer, a site investigation contractor and, where appropriate, a ground treatment or remediation contractor. There are major differences between large-scale civil engineering projects and small building developments, including the size of the budgets, the time-scales, and the expertise of those engaged in the design and construction of the developments.

While the risk management process includes various forms of risk, including technical risks, financial risks and risk perception, all types of risk have financial significance. Hazards for the human population and for the natural and built environment can result in situations where there are risks to wealth and profit for developer, investigator, designer, builder, owner, occupier, insurer and finance institution.

Different hazards are likely to occur during the construction cycle: for example, land purchase, investigation, construction, sale and re-sale. Such hazards should be examined for their impact on, for example, a housing development, in such terms as:

- cost;
- delay;
- reputation;
- householder well-being;
- re-sale value.

With whom these risks lie should be considered, bearing in mind that different risks may lie with different parties:

- developer;
- designer;
- housebuilder;
- householder;
- regulator;
- insurer/guarantor.

The new standard land condition record (LCR) (Box 6) is aimed at assisting urban regeneration. It should provide a reliable record of the physical and chemical nature of any contamination and provide details of the previous use of the land. It will help to focus the attention of developers and planners on the probable hazards of the site at the earliest possible stage of redevelopment. It has been welcomed by industry and is well supported by industry bodies such as the NHBC and the House Builders Federation. The LCR is a voluntary scheme; many of the potential beneficiaries, such as medium to small developers and landowners, are currently unaware of the initiative and its benefits and, until it becomes better known and appreciated, it is unlikely to fulfil its full potential. A register of land condition specialists, which has been established to promote a consistent approach to assessing contaminated land, is being managed by the Institute of Environmental Management and Assessment (Bennett, 2002).

A risk-based approach to housing development on contaminated land has been prepared by the Environment Agency and NHBC (2000) based on model procedures for the management of contaminated land being prepared by DEFRA/Environment Agency. Box 7 outlines the underlying model procedure stages and Box 8 the steps in the guidance, which are concerned with the identification of contaminants and are based on the source-pathway-receptor concept. Information on the above procedures and guidance has been included in this report, although this report is not primarily concerned with the effect of contamination on the human population, because a somewhat similar, although simpler, approach can be used for ground-related risks to the built environment.

Box 8 Steps in the guidance for building development on contaminated land

The Environment Agency and NHBC (2000) guidance for the safe development of housing on land affected by contamination has translated the model procedures into a series of eleven practical steps which cover all the activities from initial enquiry to completion of remedial work.

1. Establish former uses of the site. Collect physical data and undertake walk-over survey. Consult regulatory authorities.
2. Identify contaminants of concern. List industries identified in (1), industry-specific contaminants and geologically-based contaminants.
3. Develop conceptual model of the site.
4. Undertake hazard assessment. Review data and conduct exploratory investigations if further information is required.
5. Design and implement site investigation. Update (3).
6. Undertake risk estimation. Obtain generic assessment criteria or calculate site-specific criteria.
7. Undertake risk evaluation. Identify unacceptable risks from comparison of measured concentrations with appropriate criteria.
8. Identify and evaluate options for remedial treatment based on risk management objectives.
9. Select preferred remedial strategy and submit for approval.
10. Design and implement remedial works. Undertake verification of remedial action.
11. Implement monitoring and maintenance programmes. Complete project.

In the first four steps, which form the first phase of risk assessment, contaminants are identified and assessed, and a conceptual model of the site is developed. The next three steps form the second phase of risk assessment in which a ground investigation is carried out and risks are estimated and evaluated. In subsequent steps, remedial options are identified and evaluated and a remedial strategy is selected. Finally, remedial works are implemented, and monitoring and maintenance programmes are put in place.

Five particular aspects of risk management for ground-related hazards for the built environment are examined:

- Hazard identification (Part 3.4);
- Risk assessment (Part 3.5);
- Risk mitigation (Part 3.6);
- Ground treatment (Part 3.7);
- Foundations and services design (Part 3.8).

Figure 8 indicates how these various aspects inter-relate in a schematic layout of the proposed approach to risk management.

Before examining the various aspects of risk management, some consideration is given to two other important matters; firstly, the influence of perception on the tolerability of risk (Part 3.2) and, secondly, the development of a conceptual model of ground conditions (Part 3.3).

Figure 8 Risk management for building development on brownfield sites

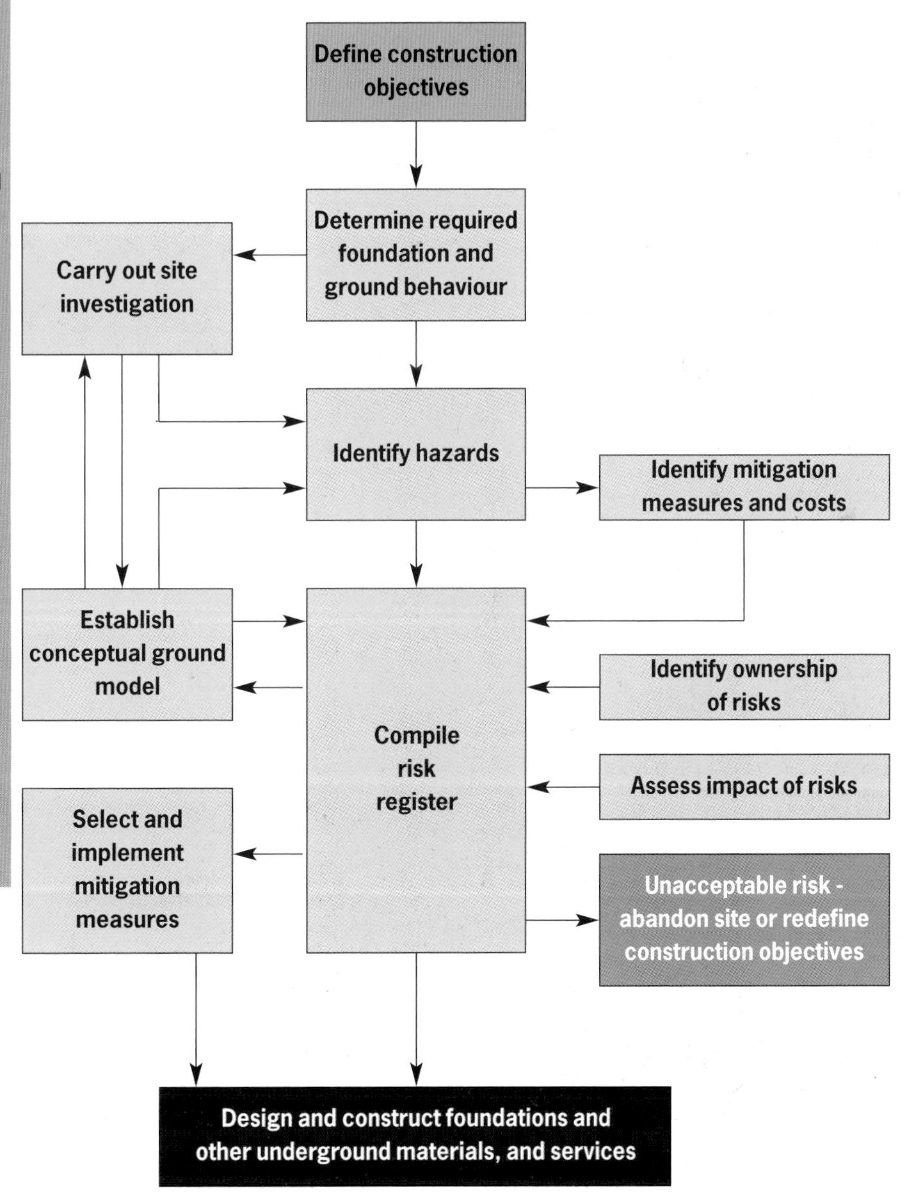

3.2 Perception and reality

Risk assessment involves an evaluation of risk and some consideration of what risk is tolerable, this latter issue bringing in the difficult subject of the perception of risk. There have been a number of studies of how people view risks (for example, Royal Society, 1992; Health and Safety Executive, 2001). It should be recognised that popular perception of risk can differ considerably from judgements that are based on scientific estimates of probability (Department of the Environment, Transport and the Regions, 2000c). This has important implications for risk management – Box 9.

Box 9 Risk: perception against reality

Factors which increase public concern include:

- risk which is involuntarily imposed;
- unfamiliar and technological risks;
- risk of a single large-scale catastrophe;
- risks which are delayed and affect children and future generations.

Clearly such perceptions cannot be ignored in risk management. It has become fashionable to assert that perception is reality, but in some situations this axiom can be dangerously misleading. It is true that actions may be governed by perception rather than by reality, and hence perception is very important. However, where perception markedly diverges from reality, actions may be based on erroneous perceptions and lead to difficulties. The difference between perception and reality is at the heart of many of the difficulties encountered in developing brownfield sites. It can be difficult to bring perception into line with reality.

In compiling the database of problems and incidents for this project, it became clear that builders and developers are principally concerned about contamination. This is understandable because human health is an emotive issue and there can be strong regulatory pressure. However, it also became clear that most ground-related problems experienced on brownfield sites to date relate to the built environment and are concerned with ground movement – Figure 9.

An account of the removal of an old landfill in a semi-rural area of Cheshire provides an insight into the problems that can be caused by local residents' perception of risk (Taunton and Adams, 2001). Although the remediation work would remove the potential hazard of a contaminated site from the locality, the perception grew among some residents that there was a potential health problem related to the work. Once such a perception is established in the minds of the residents '*it becomes reality for them*' and only by giving residents greater involvement in the project could the perception be changed.

Guidance on communicating an understanding of contaminated land risks has been provided by SNIFFER (1999). Advice is given on when, with whom, what, and how to communicate.

The perceived risks of ground contamination can act as a barrier to development. The developer becomes focused on environmental issues as a result of the difficulties in satisfying regulatory planning and control measures for contaminated land. Geotechnical problems may be poorly investigated because much of the budget has been spent on environmental investigations.

Figure 9 Perception versus reality

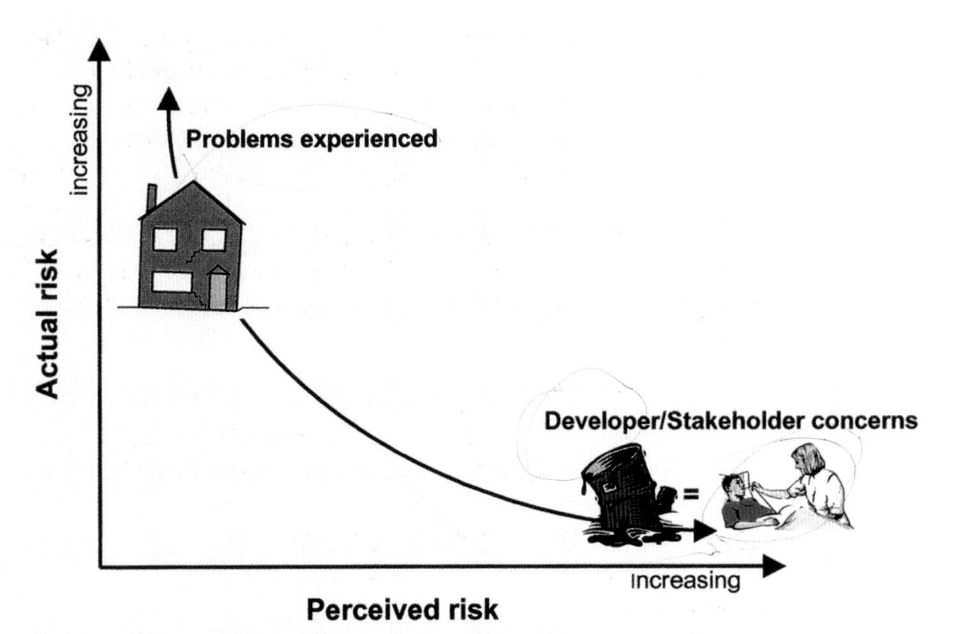

A change in philosophy is needed for geotechnical site investigation of brownfield sites. The fundamental issues must be addressed correctly and more time needs to be given to identifying likely geotechnical hazards (Clayton 2001a and 2001b). Thorough and careful desk studies and walk-over surveys are an essential step in this process. This then can be followed by a better targeted and focused ground investigation.

3.3 Conceptual model of ground conditions

The objective of a site investigation is to ensure economic design and construction by reducing to an acceptable level the uncertainties and risks that the ground poses to the project (Institution of Civil Engineers Site Investigation Steering Group, 1993). The site investigation should identify all the ground conditions which are relevant to the development. An appropriately targeted site investigation should minimise the risk posed by ground-related hazards.

The site investigation should provide the information required to form a conceptual model of the ground conditions.

- The development of a conceptual model is strongly advocated in BS 10175:2001, the code of practice for the investigation of potentially contaminated sites. Indeed, it is stated that the purpose of a site investigation is to gather the information needed to form such a model in order to be in a position to assess the presence and significance of contamination of land.
- Geotechnical engineers also develop a conceptual model of the physical ground conditions, although they may not always recognise it as such.

The principle of defining and revising a conceptual ground model for the site is a key theme and has the following elements:

- Three-dimensional stratigraphic model of the ground. The soil or soil and rock profiles need to be determined. The measurement of the engineering properties of the different strata which are relevant to foundation performance may be required.
- Modifications caused by previous human activities. An understanding of the history of the site and of the ways in which it has been affected by various types of human activity is required.
- Groundwater model. Many ground-related problems are associated with groundwater and an understanding of groundwater conditions is essential.
- Soil-structure interaction model. The physical behaviour of the ground under foundation loading and the response of the building to ground movements caused by foundation loading or other phenomena should be evaluated.
- Soil contamination model. Source-pathway-receptor scenarios should be identified (Environment Agency and NHBC, 2000).

A conceptual model of the site is enhanced as information is gathered. The site investigation may be multi-staged. A helpful procedural flowchart indicating the stages at which further investigation may be required is given in the NHBC standard: *Land quality - managing ground conditions* (National House Building Council, 1999) and is shown in Figure 10. At an early stage in the investigation, the developing conceptual ground model should facilitate the identification of likely hazards. The later stages of the investigation can concentrate on these hazards and evaluating the risks that they pose. The conceptual model should include assessing the consequences and effects of site activities.

More detailed ground investigation may include:

- laboratory tests - BS1377: Parts 1-8 provides test specifications and the Association of Geotechnical and Geoenvironmental Specialists (1998c) has provided a guide to the selection of tests;
- in-situ tests - BS1377: Part 9 provides test specifications;
- load tests - BS1377: Part 9 provides test specifications;
- geophysical tests;
- monitoring.

An intrusive investigation on a brownfield site using trial pits and boreholes can present hazards for site operatives and also poor backfilling of trial pits and boreholes can create hazards. Boreholes could permit contaminants to penetrate deeper into the ground and cause contamination of an aquifer. Trial pits can result in the intermixing of clean and contaminated ground and can also lead to soft spots due to inadequately compacted backfilling. Careful reinstatement is essential.

A wide background to site investigation including professional liabilities and insurance issues is provided in a practical handbook on site selection and investigation by Lampert and Woodley (1991) and in a book on risk management for construction professionals by Hatem (1998). Uff and Clayton (1986) have made recommendations for the procurement of ground investigation. A general account of site investigation practice is in Clayton et al (1995) and authoritative information on particular matters is contained in reports and codes prepared by BSI, CEN, ICE and AGS – see Box 10.

Site investigation for low-rise buildings presents particular problems owing to the small scale of many developments and the sometimes inadequate geotechnical input. BRE has issued Digests giving guidance on best practice for site investigation for low-rise building – see Box 11.

Box 11 BRE Digests on site investigation for low-rise building

322	Procurement
318	Desk studies
348	The walk-over survey
381	Trial pits
383	Soil description
411	Direct investigations

Box 10 Codes and reports on site investigation

British Standards Institution
- BS 5930:1999 Code of practice for site investigations.
- BS 10175:2001 Code of practice for investigation of potentially contaminated sites.
- DD ENV 1997 Eurocode 7. Geotechnical design
  DD ENV 1997-2: 2000 Design assisted by laboratory testing.
  DD ENV 1997-3: 2000 Design assisted by field testing.

Institution of Civil Engineers
- Site investigation in construction. Site Investigation Steering Group, 1993.

Association of Geotechnical and Geoenvironmental Specialists
- Code of conduct for site investigation, 1998.
- Guidelines for good practice in site investigation, 1998.
- Guidelines for combined geoenvironmental and geotechnical investigations, 2000.

National House Building Council
- NHBC Standards Chapter 4.1. Land quality – managing ground conditions, 1999.

Department of Environment (now DEFRA)
- CLR 4, Sampling strategies for contaminated land. Nottingham Trent University, 1994.

Figure 10 Flowchart for managing ground conditions *(NHBC 1999)*

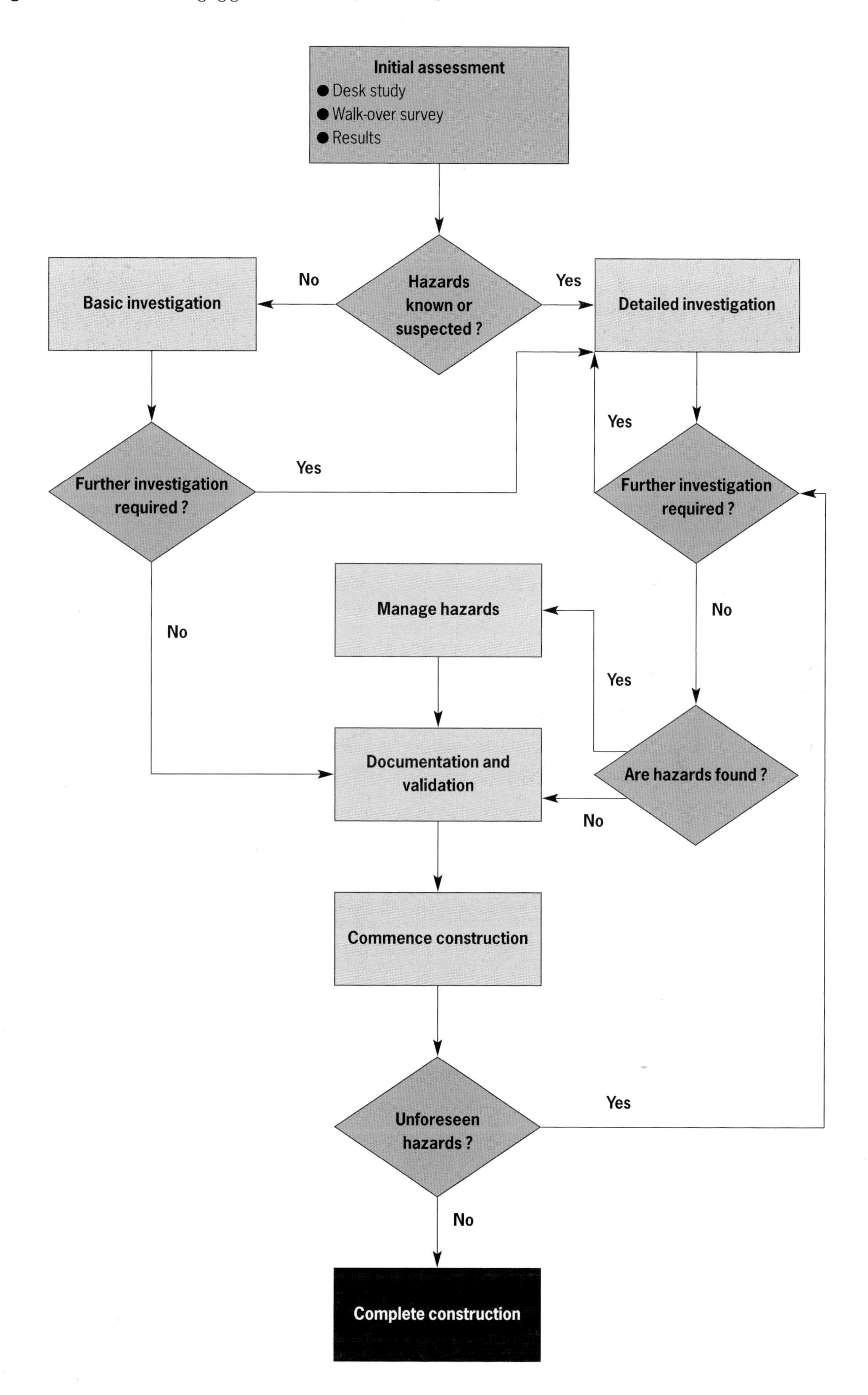

3.4 Identifying hazards

It is important to identify hazards at an early stage. Case histories 1 to 4 illustrate the difficulties that can arise where hazards are identified too late:
- During construction, more contamination was found than had been predicted from the site investigation and further investigations and monitoring were required (case history 1).
- At three plots, old buried foundations at shallow depth prevented necessary piling being carried out (case history 2).
- Construction over the edge of a backfilled quarry resulted in subsidence of the property (case history 3).
- The extent of filled ground was not correctly identified by the site investigation and more properties required piling than had been anticipated (case history 4).

Box 12 Settlement of houses on opencast mining backfill (Thompson, 1998)

Vibro stone columns were installed beneath the house foundations at a housing development on a 10 m-deep clay and mudstone opencast backfill in the West Midlands. Shortly after the development had been completed, some of the properties began to suffer total and differential settlement. The houses were built on reinforced concrete raft foundations and experienced little cracking and damage. Nevertheless, over the 9 m width of a detached house, tilts reached 300 mm. The costly remedial works involved the installation of piles and jacking the houses back to level. Thompson (1998) concluded that the vibro stone columns would have provided a means of introducing water into the backfill, encouraging collapse compression.

Box 12 shows an example of the serious costs which may be incurred if the ground conditions at a site are not adequately understood at an early stage and the significance of a major hazard is not fully appreciated . Thompson (1998) concluded that '... *if changes are found to be necessary at the operation stage, then the cost is likely to be far higher than if the condition of the site is understood at an earlier stage such as during design. Indeed when substantial and costly changes are required, it is possible that litigation proceedings may result adding further expenditure. It is clearly important for geotechnical engineers to ensure that they are in a position to provide appropriate specialist advice in a timely manner so that minimal changes are found to be necessary to the below ground works once the construction stage is complete*'.

On a typical brownfield site, there is a wide range of possible geotechnical and geoenvironmental problems and it is important at an early stage to identify the principal likely hazards and focus the site investigation on these. This will typically involve the following stages – see Figure 11.
(1) appoint specialist adviser;
(2) desk study;
(3) site reconnaissance with walk-over survey;
(4) collate information and apply expert knowledge to develop a conceptual soil model;
(5) identify hazards and consequent risks to built environment.

(A) Ground movement

Most types of ground will be subject to movement owing to applied loads, moisture movement and other effects. Ground movements commonly encountered at brownfield sites are described in Part 2.2 and the question which has to be addressed is whether the form, magnitude and duration of movement will significantly affect building development. Monitoring the ground movements will establish the current situation but will not identify those situations where there is a potential for large ground movements due to some environmental change (for example collapse compression of fill when the moisture content is increased).

Compressible fills

It is important to establish accurately the extent, depth and content of filled areas. Abrupt changes in depth of fill are the most likely locations for damaging differential settlement and such locations are not necessarily restricted to the edges of filled areas. Information obtained from old records and plans should be confirmed by trial pits and boreholes.

Disturbed samples may give a good indication of the nature of the fill, but for many types of fill it is difficult to obtain reasonably undisturbed samples which are needed to assess the density of the fill. In-situ penetration tests, such as dynamic probing and the standard penetration test, can be used to assess the condition of those fills which do not contain very large particles. These tests may also give some indication of the density of the fill and so provide additional information. Appropriate geophysical tests may also help in some situations – see CIRIA report C562 (McDowell et al, 2002).

In view of the significance of collapse settlement for building development, the quantification of collapse potential is a key element in an investigation of a filled site prior to building. All available information should be used in the assessment, including historical information and site investigation data. With adequate knowledge of the origin and subsequent history of a fill, some preliminary judgement as to whether collapse compression is a hazard can be made (Charles and Watts, 1996; BRE Report 424; Charles and Skinner, 2002). It should be assumed that any partially saturated fill placed without systematic compaction, including poorly reinstated trial pits, is vulnerable to collapse compression on wetting unless there is adequate evidence to the contrary, such as knowledge that the fill has been submerged some time in the past.

There may also be significant potential for volumetric compression due to creep or the application of load.

Expansive fills

Large deposits of mixed slags and other wastes are usually associated with iron and steel-making plants and sites in the same locality. Where these plants have been operating comparatively recently, the main slag deposits should be

Figure 11 Identifying ground-related hazards

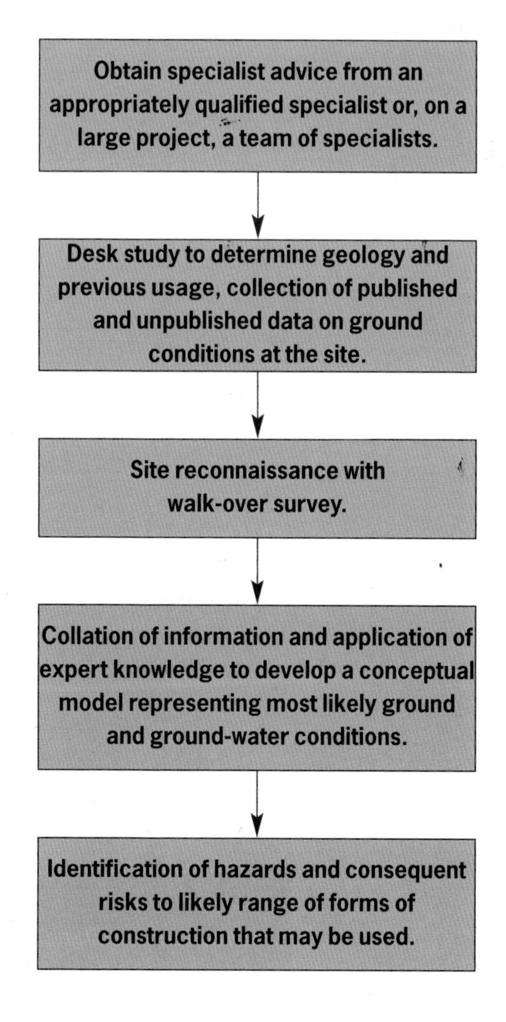

readily recognised and located. Many industries, including some car manufacturers, operated their own foundries, so slags may be found on sites not directly related to the iron and steel industry. Slags may be present as a result of disposal as waste or use as fill or hardcore.

Procedures for assessing the stability of slags have existed for many years. For air-cooled blastfurnace slags, for use as aggregates, these are set out in BS 1047. For steelmaking slags, a variety of tests have been used to detect the potential for expansion and to determine its likely magnitude, although there is no British or European Standard method at present.

The procedures for on-site sampling and laboratory preparation and sub-sampling are critical factors in obtaining reliable and representative assessments of slag stability. It is usually possible to identify a potential to expand but it is questionable whether the magnitude of expansion in the field can be predicted from laboratory test results.

Biodegradable fills
It is important to identify where there is a substantial volume of putrescible waste because there will be a potential for major reduction in volume due to biodegradation. It can cause large ground movement and gas generation.

Buried foundations, pipework and storage tanks
The location of old foundations and services may be determined from old records and should be confirmed by trial pits. It is easy to overlook this type of buried hazard – see case history 2.

Shallow mine-workings
The location of old mine-workings may be determined from old records and confirmed by boreholes. In an area where shallow mine-workings are expected, but have not been located, the likelihood of them being present needs to be assessed (Healy and Head, 1984). Where shallow mine-workings have been identified, the likelihood of their collapse should be examined. The Law Society (1998) and the Law Society of Scotland (1999) have published guidance notes and a directory for coal mining searches.

(B) Vulnerability of construction materials to aggressive ground conditions

Aggressive ground conditions can occur on brownfield sites and are often associated with the presence of contaminants. A key issue is whether or not this presents any problem for the proposed development and whether any action is required. It is necessary to determine the type, nature and concentration of potentially hazardous substances. Soils and groundwater can be sampled using trial pits and boreholes following the principles of good practice described in BS 10175.

Chemical attack below ground is difficult to assess owing to poor accessibility. Damage has generally been found on the tops of exposed piles or foundation walls but is unlikely to be confined to the top area. The form of concrete is important in determining the risks of attack. Modern housing, which tends to be of lighter construction, should be carefully designed and specified. Some services are particularly vulnerable to aggressive ground conditions. For example, unprotected MDPE water pipe is susceptible to organic attack and permeation by hydrocarbons as described by Garvin et al (1999).

Many factors influence deterioration and, if it does take place, the rate of deterioration. Identification and management of these factors, which are listed in Part 2.3, are part of the risk management process.

(C) Gas migration

Most problems concerned with the migration of gas within or from the ground are at or adjacent to landfill sites. There is, therefore, a potential problem where biodegradable material is identified. Identification of migration pathways through the natural ground surrounding the landfill requires a good understanding of the geology.

Guidance for good practice in the design and execution of site investigations for methane and associated gases in the ground has been provided by Raybould et al (1995). Gas analysis and monitoring (BRE Report BR 100; Crowhurst and Manchester, 1993) can confirm the existence and magnitude of the hazard. However, valid interpretation depends on reliable measurements and Harries et al (1995) have given guidance on testing the validity of gas measurements.

There are major implications where gas is being generated in significant quantities, not only for hazards to property but also for health and safety. Approved Document C of the Building Regulations (Department of Transport, Environment and the Regions, 2000d) gives the general approach to building development on land with gaseous contaminants such as landfill gas and methane – see Part 2.4.

Card (1995) has examined the need to protect buildings from hazards associated with methane and other gases, and O'Riordan and Milloy (1995) have presented a rational methodology for gas hazard evaluation and risk assessment. It provides an example of the use of quantitative risk assessment in the brownfield area.

(D) Subterranean fires

Subterranean fires have occurred in such combustible fills as colliery spoil and domestic refuse. Site inspections can reveal tell-tale signs of fires, which include visible signs such as smoke plumes and settlement. Where thermal activity is suspected, ground temperature surveys should be carried out (Bedford and Smith, 1988). The sub-surface temperature may be determined by instruments lowered into boreholes or by probes driven into the ground. In boreholes, air or water movement may affect the measurements.

Subterranean fires require the presence of combustible material, a source of ignition and a flow of oxygen. Where a site has been identified as containing potentially combustible material, assessment tests can be undertaken on samples of ground materials. However, these tests are limited in their use and the results difficult to relate directly to site conditions. Some guidance is given in ICRCL 61/84. The guidance on assessing ground combustibility is dated, and the tests for assessing combustibility do not provide an adequate indicator of actual site conditions: how susceptible ground material is to ignition from either self-heating or external sources, such as a fire on the ground surface.

3.5 Assessing risk

Having identified likely hazards, the risks they pose for the proposed development should be assessed. Risk assessment involves exploring the circumstances in which hazards could give rise to unacceptable consequences. The following elements can be identified:

- the identification and cataloguing of hazards;
- the estimation of the probability of occurrence and the estimation of its consequences;
- the ranking of risks and categorisation in terms of actions required;
- the tolerability of the consequences.

The risks should be ranked according to their impact and the different phases of the project with which they are associated.

Box 13 Risk register

A risk register provides a means of recording data and decisions so that information on risks can be effectively communicated. A risk register may include the following:

- Identified hazards.
- Nature and degree of risk resulting from the hazards.
- Planned response.
- Estimated effect of response.
- Nature and degree of residual risk and with whom it lies.

A risk register should be established – see Box 13. The use of a risk register for managing geotechnical risk has been described by Clayton (2001a and 2001b) and for geoenvironmental investigations by Summersgill (1997). Risk registers are also used in managing health and safety risks (Health and Safety Executive, 1998 and 1999). A risk register provides a means of systematically recording data and decisions, so that information on risks can be readily communicated within organisations and between the different parties involved in the building development.

In most cases it will not be feasible to calculate the probability of occurrence of a particular hazard with any degree of certainty. Some simple qualitative scales for estimating the degree of risk are needed. Table 2 gives a conceptual matrix for the estimation of the degree of risk from a consideration of probability and consequences.

The simple classification of risk in Table 2 can be used to identify high risks and so assist in focusing resources on areas where they are most needed and in targeting particular areas of concern. The risk increases, and the acceptability of the hazard decreases, moving from left to right across the table and upwards.

- Where the consequences are negligible, the degree of risk is negligible, irrespective of the probability of occurrence.
- Where the probability of occurrence is high and the consequences are severe, the degree of risk is high.
- Where the probability of occurrence appears to be almost negligible but the consequences are severe, the risk is difficult to evaluate.

There is a major difficulty where the probability of occurrence appears to be almost negligible but the consequences are very severe. In an extreme case, assessing the degree of risk involves multiplying a probability which is close to zero by consequences which may seem to approach infinity! In this situation, the table correctly (but not very helpfully) indicates that the degree of risk could be low, medium or high. Fortunately, this type of situation is not normally found in the circumstances which are the subject of this report, but may be encountered where human health is at risk – shown diagrammatically in Figure 12.

Table 2 Degree of risk *(Department of the Environment, Transport and the Regions, 2000c)*

Probability	Consequences			
	Negligible	Mild	Moderate	Severe
High	Near zero	Medium/low	High	High
Medium	Near zero	Low	Medium	High
Low	Near zero	Low	Medium/low	High/medium
Negligible	Near zero	Low	Medium/low	High/medium/low

Clayton (2001b) has presented some simple numerical scales for probability and consequences in relation to geotechnical risk for construction costs. The degree of risk can be computed by multiplying the likelihood and the effect. The product of these two assessments can then be judged against another qualitative scale. However, the approach assumes an equivalence between probability and consequence and ignores the difficulty which arises where the probability of occurrence appears to be almost negligible but the consequences are very severe. Furthermore, whole life costs need to be considered, not just construction costs. Tolerance to risk varies widely depending on the business or individual taking the risk.

Despite its limitations, the type of approach shown in Table 2 provides a useful way of comparing degrees of risk and prioritising the need for risk mitigation.

3.6 Mitigating risk

Having identified the likely hazards and assessed the risks that they pose for the proposed development, appropriate action must be taken to avoid, prevent, mitigate or transfer risk. The risks should be ranked according to their impact and the different phases of the project with which they are associated. As described in Part 3.5, a risk register should be established and likely costs assessed.

The value of risk assessment is in developing a risk management strategy; where the risks posed by a particular hazard to a building development are unacceptable, appropriate action should be taken. Once the hazards are identified and the risks they pose to the building development are assessed, it should be possible to select and implement a technically adequate solution. However, a solution to one problem may cause some other difficulty. For example, piles or vibrated stone columns installed to reduce settlement may provide a pathway for contaminants to move through the ground (Westcott et al, 2001a and 2001b) (Parts 3.7 and 3.8).

Figure 12 Degree of risk

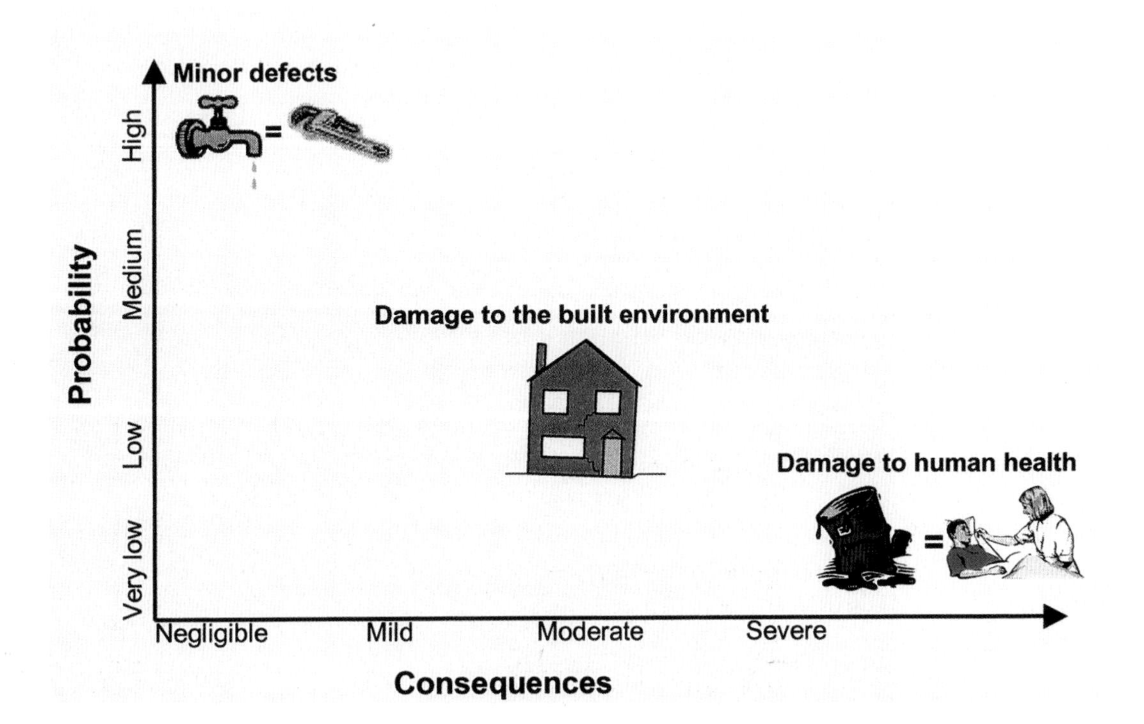

The selection of an appropriate remedial strategy involves many factors related to technical adequacy, costs, environmental effects and perception.

- **Technical adequacy** It is essential that the problem has been diagnosed correctly and a technically adequate solution is adopted. Any other factors must be subservient to this over-riding imperative.
- **Costs** A strategy which involves relocating the development to another site may have large financial implications, whereas a decision to mitigate the problems of differential settlement by applying some form of ground treatment and/or providing stiff raft foundations may add relatively little to the cost of the development.
- **Environmental effects** A solution that is technically adequate for dealing with the hazards that have been identified could have deleterious environmental effects.
- **Perception** Where a site is popularly perceived to be hazardous, it is important to adopt a strategy that will yield a solution which is transparent.

The following are some basic types of remedial strategy.

Avoid creating the hazard

There are situations in which it may be possible to avoid creating a particular hazard. For example, where the site reclamation process has not yet been carried out, the earthmoving operation can be controlled so that fill materials which can cause both expansive reactions and the generation of highly polluting leachates are not mixed. The extra costs will be small and the gain may be considerable in terms of risk mitigation.

Placement under controlled conditions of an adequately engineered fill (Figure 13) can eliminate or greatly reduce the hazard of compressible fill and many other hazards described here (Trenter and Charles, 1996; Charles et al, 1998). The extra costs involved in adequate compaction of the fill and suitable supervision will not normally be large. The consequences of failure can be great in terms of building reconstruction and expensive legal action.

Figure 13 Controlled placement of fill

Relocate the building development

Where major hazards are identified and the risks are great, abandonment of the site or part of site for the intended purpose should be considered. On large sites it may be possible to avoid using the worst areas for the most sensitive end uses. This type of approach can be very helpful on large developments which include residential, retail and commercial buildings on sites containing many different types of brownfield land. There will be much less scope on smaller sites. An unwarranted over-sensitivity to risk will defeat the objective of locating building developments on brownfield sites.

Segregate the buildings from the hazard

The foundations of a building can be isolated from the effects of a compressible fill by the use of piles which transmit the weight of the structure to a firm stratum underlying the fill. In case history 3, where this approach was not successful, it is believed that the original piles did not reach the quarry floor. Even where the building is successfully isolated from a settling fill, services will be subjected to severe differential settlement where they enter the building as the fill settles around the building.

Containment of contamination has a vital role in the utilisation of many brownfield sites. In this commonly used remediation method, the objective is to block or control the pathway between a hazard and potential receptors. Engineered cover layers and in-ground vertical barriers, such as slurry trench cut-off walls (Box 14), have been widely used as reliable means of containing contamination. General guidance on the use of cover systems is given in CIRIA reports SP 124 (Privett et al, 1996) and SP 106 (Harris et al, 1995). Where the hazard is associated with the migration of gas, the flow of groundwater or the spreading of a subterranean fire, the site may be protected by a cut-off wall. A specification for slurry trench cut-off walls has been prepared by the Institution of Civil Engineers (1999).

Box 14 Slurry trench cut-off walls – *see also Digest 395*

Slurry trench cut-off walls using self-hardening cement-bentonite are the most common type of vertical barrier used to control lateral migration of pollution and gas from contaminated land and landfill sites in the United Kingdom. To form a satisfactory barrier to leakage migration, a cut-off wall must have a low permeability and an adequate toe-in to an underlying aquiclude. Despite extensive use for waste and pollution management, little is known about either the groundwater contaminant conditions under which a slurry wall will remain intact without increase in permeability or the capacity of a slurry wall to treat contaminants as they migrate through it. These uncertainties lead to difficulties in guaranteeing long-term retention of contaminants inside slurry-wall enclosed sites (Tedd et al, 1997). Laboratory tests have shown that deterioration of slurry walls could occur in the long-term in very chemically aggressive ground conditions, but field experiments have shown that even in very aggressive ground conditions little deterioration takes place (Tedd, 2001). For control of gas migration, it is normally recommended that the slurry wall includes an HDPE membrane.

Remove the poor ground

This solution involves the physical removal of the problem soils to another site and, if necessary, placement of an imported engineered fill; removal off-site may involve major costs. Alternatively, following excavation of the poor ground, it may be sorted to facilitate removal of unacceptable material and the remaining acceptable material can be placed as an engineered fill.

Improve the poor ground in-situ

Some form of in-situ ground treatment can be used to improve the physical or chemical properties of the ground so that the ground-related risks to the development are mitigated. Ground treatment is described in Part 3.7.

Design foundations and services to cope with the hazard

The foundations can be designed to withstand the aggressive conditions and protect the building from the ground-related hazards. Foundation design is described in Part 3.8.

3.7 Ground treatment

One approach to risk mitigation is to apply some form of in-situ ground treatment or remedial process prior to building development; on many sites ground remediation work will be required. There are many different types of hazard and there are many different solutions. Decisions on how to manage physical and chemical hazards in brownfield sites depend critically on accepted criteria for performance which is suitable for use and there is some uncertainty about these criteria.

In determining an appropriate risk mitigation strategy, it will be necessary to assess the cost of the proposed measures and the extent to which they will reduce the risks to which the development is exposed. Some forms of ground treatment have a generally beneficial effect with regard to several hazards, whereas other methods may help with one hazard but introduce other potential problems.

- Densification of fill using some form of compaction technique that is applied at the ground surface will generally reduce the potential for subsequent settlement of the fill. It may also inhibit groundwater flow and gas migration and reduce the likelihood of combustion.
- Installing vibrated stone columns may reduce the compressibility of the ground but highly permeable pathways may have been introduced that could cause additional difficulties (Westcott et al, 2001a and 2001b).

In many cases ground treatment will substantially reduce the ground-related risks to the building development for expenditure that is small compared to the overall cost of the development. In addition to improving the properties of the ground, the treatment can in effect act, when appropriately monitored, as an auxiliary site investigation.

(A) Ground movement

Many of the problems on brownfield sites are associated with ground movement and it is appropriate to first consider treatment techniques that can reduce the potential for such movements.

Compressible fills

Where the load carrying characteristics of the ground are inadequate owing to the presence of compressible fills, there are several widely used methods of ground treatment which will reduce the compressibility of fills, including vibrated stone columns (vibro), dynamic compaction and pre-loading. BRE has carried out major studies of the effectiveness of these various treatment methods (BRE Report BR 424; Watts and Charles, 1997).

Whereas dynamic compaction and pre-loading are applied at the ground surface, vibrated stone columns, currently the most common form of ground treatment on brownfield sites, is a penetrative treatment technique. Figure 14 shows the installation of a vibrated stone column. *Specifying vibro stone columns* (BR391) gives guidance on the type of stone to be used and makes particular reference to the unsuitability of limestone in acid ground conditions and to due regard being given to the presence of any chemical contaminants which have been identified. Long-term degradation of the stone which forms an integral part of the foundation system could result in unacceptable movements of the building. Westcott et al (2001a and 2001b) have examined the potential hazard of piling and penetrative ground improvement methods creating pathways by which contamination can affect controlled waters, but material degradation is also an issue.

The cost of ground treatment depends on the size of the area being treated and whether treatment can be confined to the ground directly below the foundations or whether it is necessary to treat a much larger area. Despite these complications, it is helpful to have some idea of typical costs. Dynamic compaction and preloading are appropriate for treating large areas whereas vibrated stone columns can be located under the strip footings of low-rise housing. Because of the size of the crane which is required for dynamic compaction, the mobilisation costs are large and it is unlikely to be economic on areas smaller than 10 000 m^2. Preloading with a surcharge of fill is also unlikely to be practicable on a small site and the cost will depend on the haul distance for the fill used as a temporary surcharge. Ground treatment for a small detached house might typically be as little as £2000 for a simple, relatively shallow, vibrated stone column treatment and as much as £5000 for a substantial preloading.

Expansive fills

Great care is needed during reclamation of industrial complexes to avoid inadvertently mixing blastfurnace or steel slag with other fills, such as colliery spoil or domestic refuse, because this can cause both expansive reactions and the generation of highly polluting leachates. This can occur if waste fills are excavated and then replaced as an engineered compacted fill. Care is also needed when importing fill to ensure that it is not contaminated.

Figure 14 Installing vibrated stone columns *Pennine Vibropiling Ltd*

Biodegradable fills
Buildings can be designed to be protected against the ingress of methane and associated gases (described later) but it is advisable, prior to building, to remove fills with a large biodegradable content or relocate the development.

Buried foundations and pipework
It is preferable to excavate buried foundations and pipework, not only to remove obstructions to construction of the foundations but also to prevent local hardspots causing differential settlement. Alternatively, new foundations may need to be designed to take account of existing hardspots.

Old mine-workings
Where there is a risk of collapse of old mine-workings, ground treatment involves filling up the old workings. The most common solution is to fill the workings with a pfa:cement-based grout (Sear, 2001) via boreholes. The factors that control the spacing of the boreholes are described by Healy and Head (1984); Atkinson (1993) comments that the boreholes are generally located on a 3 m grid. Where the seam is thick and there are large voids, it may be necessary to install a grout curtain or other type of barrier around the perimeter of the ground to be treated. This type of treatment can involve major expense. Low cost pfa pastes have been developed for large abandoned limestone workings (Jarvis and Brooks, 1996).

(B) Vulnerability of construction materials to aggressive ground conditions
For chemically contaminated ground, a key issue is whether or not any action is required. In many cases no action is taken other than to design foundations and services which will be adequately resistant to the aggressive environment (Part 3.8). Appropriate concrete design and placing services in clean trenches is often adequate. Where some remedial action is considered to be necessary, there are three principal alternatives:
- remove material ('dig-and-dump');
- contain material in the ground;
- treat the ground to remove the contamination.

The majority of contaminated land in the UK is remediated by excavating and transporting to contained landfill or, more directly, by on-site containment. There may be a preference for removal of the contamination off-site due to concerns over future liabilities but this option may become increasingly financially unattractive as measures are taken to limit environmental impact, both in transit and at the receiving site.

Fill which contains materials hostile to the built environment should be removed from locations where groundwater may rise to the underside of foundations and should not be permitted within a depth of 1 m from the underside of the foundations. This includes natural materials such as pyritic shales and gypsiferous clays, and waste materials such as burnt colliery spoil and steel slag.

(C) Gas migration
Barriers, including cut-off walls and cover layers, and vent trenches can be constructed in or on the ground to prevent the migration of gases (Card, 1995). Slurry trench cut-off walls are described in Part 3.6.

(D) Subterranean fires

Building on sites with subterranean fires should be avoided but a hazard remains where combustible material has been left in place under a building. Where a potential problem has been shown to exist, there are various preventative measures that can be implemented to ensure a fire will not start:

- remove combustible materials;
- mix inert substances with the combustible materials;
- compact the site to remove air voids;
- protect the site against possible ignition sources.

A surrounding cut-off wall may be used to form an isolation barrier to prevent migration of an underground fire into the ground under the building (Palmer, 1979; Weatherley, 1979).

Although the literature on prevention and control of subterranean fires is better than on other aspects of this subject, there is still no up-to-date published guidance. The following techniques have been used to control and extinguish subterranean fires:

- isolation barriers;
- compaction;
- smothering;
- flooding;
- injection of inert materials;
- excavation;
- controlled burn-out.

3.8 Foundations and services design

General guidance on foundation design can be found in Atkinson (1993) and Curtin et al (1994), as well as in BS 8004. Eurocode 7 specifies rules for geotechnical design. While the need for a close link between foundation design and ground-related problems might appear self-evident, it is not always present in practice. Where a building is to be constructed on a brownfield site, it is important that the geotechnical and geoenvironmental specialists understand the requirements of the structural designer and that the structural designer, in turn, has realistic expectations of ground behaviour. A robust design is required which can resist a variety of ground conditions.

Foundations are primarily required to support the building but can also act as a barrier to contaminants. Interaction with contamination could result in loss of strength or change in volume of the foundation material, which could cause damage to the building or services as they enter the building.

Foundations can be broadly divided into shallow and deep types. Deep foundations which effectively by-pass near-surface soils with poor load-carrying properties may be preferable from a structural point of view, but it may be better environmentally to have shallow foundations which do not penetrate deeply into contaminated ground.

Shallow foundations include:

- strip and trench fill;
- pad;
- raft.

Strip, trench fill and pad foundations are generally less than 1 m deep, but they can be up to 2 m deep and, very occasionally in shrinkable clay soils, can be as deep as 3 m. On many sites, contact with groundwater contamination is unlikely. Figures 15 and 16 show these different types of shallow foundations (Martin, 1996).

Figure 17 shows a typical raft design. Raft foundations are generally built onto an engineered sub-base of clean granular fill, close to ground surface and above the water table, and the availability of contaminants to the foundation is considerably less than it is for deeper foundations. The granular fill and/or impermeable membrane can be used as a capillary break within a clean cover system to control upward migration of contaminants.

Figure 15 Trench fill foundation

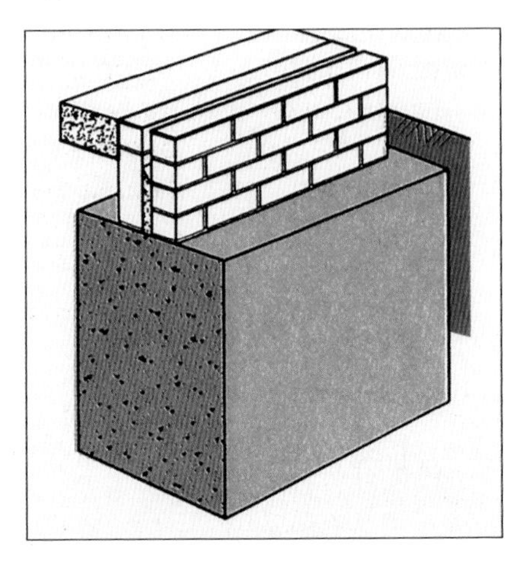

Figure 16 Pad foundation

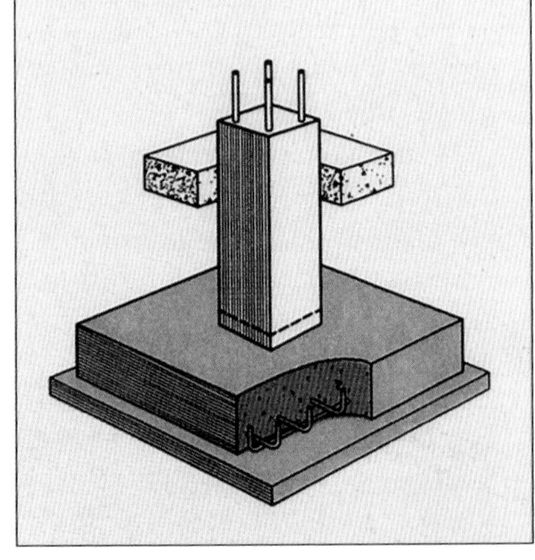

Piling and penetrative ground improvement methods, such as vibrated stone columns and vibrated concrete columns, are commonly used for providing foundations on brownfield sites. These foundation systems and ground treatment processes are more likely to pass through any contamination on site and to come into contact with any contaminated groundwater. Materials used, therefore, must be sufficiently durable to resist degradation. Westcott et al (2001a and 2001b) have reviewed piling methods on land affected by contamination.

Services in the ground include pipework carrying water, gas, electricity and telephone cables, drainage gullies, culverts, sewers and foul drainage, and soakaways for rainwater. Materials used for underground services and foundations for overground services include concrete, plastics, ceramic and metal pipes, coatings for pipes, sheathing and insulation of electric cables. Physical and chemical hazards must be considered in the design of the services in relation to matters such as protection against excessive differential settlement, the permeation of polyethylene pipes with organics, and the need to avoid existing buried obstructions.

Many brownfield sites result from building demolition and contain substantial foundations. The problems that these can cause for redevelopment, particularly as a cause of differential settlement, are described in Part 2.2; it can be a considerable deterrent to developers to have to remove or bridge over old foundations, or to site buildings away from them. The use of existing foundations could deliver substantial benefits to clients, but there are a number of potential difficulties:

- The proposed new building may not be geometrically consistent with the old foundation layout.
- The new structure may impose loads for which the old foundation does not have sufficient capacity and load distributions markedly different from the original, leading to the risk of differential movement and deformation of the new building.
- The remaining useful life of the old foundations may be inadequate; factors to be considered include the state of the materials, any degradation due to sulfate attack and the position of reinforcement.

Figures 15, 16 and 17 are reproduced by kind permission from CIRIA Special Publication 136

Figure 17 Raft foundation

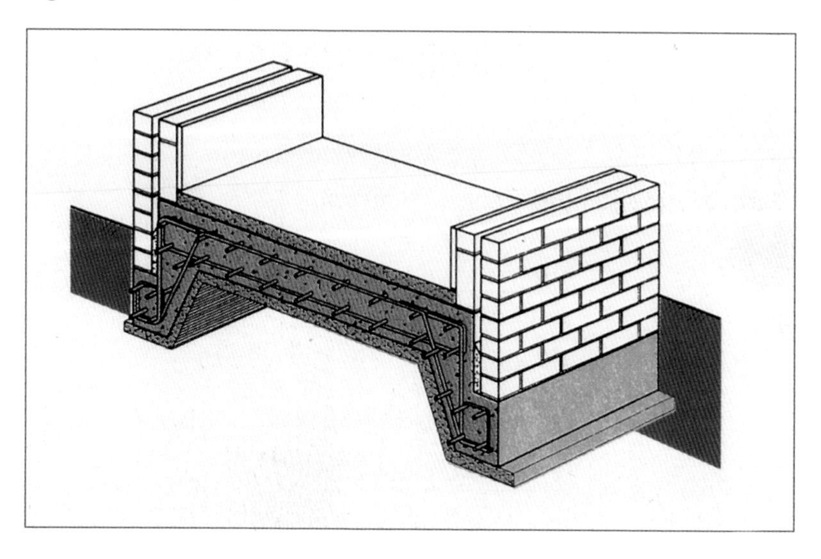

(A) Ground movement

Foundations are designed generally to minimise differential ground movements: the building design assumes that the foundation will preserve the superstructure from damaging movements. When a small building is built on compressible fill, the rigidity of a stiff reinforced concrete raft should preserve the building from damaging distortion owing to differential settlement and horizontal strains, but a risk of unacceptable tilt remains. For large buildings, this approach may be impracticable; articulated construction has been used for buildings such as schools in areas of mining subsidence but this can be expensive (Atkinson, 1993).

The design of a stiff raft for a detached house or semi-detached house is commonly based on some empirical design procedure. The raft is usually placed on a layer of granular fill. Current practice on filled sites is to design a raft with edge beams such that a distance x can be spanned and a distance $0.5x$ at the edge of the building can act as a cantilever. For fill which is old, well layered and without significant voids or organic matter, Atkinson (1993) suggested $x = 3$ m and that the edge beams should have a depth $d = 0.45$ m. However, many structural engineers use $x = 4$ m and $d = 0.6$ m on filled sites. Although this type of empirical procedure appears to have been generally successful, many of the rafts built to this type of design may not have been tested by large differential movements. The extra cost of such a foundation for a small detached house is likely to be around £3000; this is small in relation to the selling price of a house and quite negligible when compared with the financial consequences of building damage which could occur if such foundations are not provided.

Where buildings are located on shallow or medium-depth fill, poor load-carrying characteristics of the fill can be circumvented by using piled foundations and a suspended floor. Piling is not usually economic for small buildings, but where it is used there can still be problems:

- Installation problems may occur where driven piles hit obstructions.
- If the depth of fill is not accurately known at the pile location, it may be incorrectly assumed that the pile has reached natural ground when it has only hit an obstruction within the fill (case history 3).
- The piles should be designed for down-drag (negative skin friction) caused by settlement of the fill. These forces may be large compared with the weight of a low-rise building.
- Special attention is needed in the design of services that span from the filled ground into buildings founded on piles.
- In a fill in which methane gas is being generated by the decay and decomposition of organic matter, piles could form paths for the gas to escape.

(B) Vulnerability of construction materials to aggressive ground conditions

The form of construction of the building and its foundations affect the vulnerability of the building materials to chemical attack. In general, the thicker the building element, the less likelihood there is of contaminant attack causing damage to the component or serious damage to the building. Foundation elements such as piles, strip footings and rafts have different exposures to attack. Slender sections are more at risk than massive foundations. Thicker building elements will last longer because of the overdesign of the element, but durability is also related to the rate of attack. Slabs and floors on the ground are especially at risk where they can dry from the top, so encouraging the movement of contaminants into the concrete from the ground.

Concrete should be designed according to Special Digest 1. Section 1E of Approved Document A1/2 of the Building Regulations (Department of the

Environment, Transport, and the Regions, 2000d) provides rules for the construction of strip foundations for low-rise residential buildings, including the design of the concrete to be used in chemically aggressive ground.

Precautions which prevent or control chemical attack on foundations include:

- Use of suitable materials, for example, appropriate choice of cement and aggregates for mortar and concrete.
- Appropriate control of production and placing that produces a dense concrete with adequate cover to reinforcement.
- Protection from ground conditions.

(C) Gas migration

The hazards posed by gas migration should be taken into account in the design and construction of buildings. Guidance on the construction of buildings near landfill sites is in BRE Reports BR 212 and BR 414 and CIRIA Report 149 (Card, 1995). Guidance on protective measures for new dwellings with regard to radon is in BRE Report BR 211.

There are two types of approach to gas protection:

- preventing or regulating gas emissions and migration from the gassing source;
- preventing migration of gas into buildings and associated infrastructure.

Approved Document C of the Building Regulations (Department of the Environment, Transport, and the Regions, 2000d) gives the general approach to building development on land with gaseous contaminants such as radon, and landfill gas and methane. A warning is given for developments adjacent to landfill sites that where development is proposed within 250 m of the boundary of the site, further investigation should be made to determine what, if any, protective measures are necessary. The Approved Document refers to BRE guidance:

- Where the level of methane in the ground is unlikely to exceed 1% by volume and the construction of the ground floor of a house or similar small building is of suspended concrete and ventilated as described in BRE guidance, no further protection is necessary.
- The concentration of carbon dioxide must also be considered and should be judged independently of the methane concentration. A carbon dioxide concentration of greater than 1.5% by volume in the ground indicates a need to consider possible measures to prevent gas ingress. A 5% by volume level in the ground implies that specific design measures are required.
- The use of permanent continuous mechanical ventilation to ensure that methane or carbon dioxide does not accumulate in or under a house is not usually feasible since there is no management system to look after it. Passive protection is generally effective only where gas concentrations are low.

In other cases and for non-domestic buildings, expert advice should be sought. Investigation together with expert advice should be used to assess the risk posed by the gas and extended monitoring may be needed. Protective measures should be incorporated into the design of the building with the detailed assistance of experts and arrangements should be made for maintenance and monitoring.

(D) Subterranean fires

Building on sites where there are subterranean fires should be avoided. On sites where combustible material has been left in place under a building, a potential hazard remains. In such a situation, a surrounding cut-off wall may be required to prevent migration of an underground fire into the ground under the building (Weatherley, 1979).

Part 4 Case histories

Descriptions of risk-based approaches to problems may appear remote from the real world in which difficult decisions have to be made and where those decisions lead to actions that can have serious consequences. The study of case histories provides a useful antidote to excessive theorisation and to the tendency for a gulf to develop between theory and practice. Case histories where difficulties have been encountered are particularly instructive. These seven case histories illustrate some of the common problems that are encountered in building on brownfield land and the need for appropriate risk management.

There is a real risk that descriptions of risk-based approaches to problems with the inevitable accompanying schematic layouts of risk management procedures appear excessively theoretical and divorced from practice. Case histories where difficulties have been encountered are particularly instructive and the following seven case histories illustrate some of the common problems that are encountered in building on brownfield land. They are concerned principally with ground-related hazards to the built environment, but there are also environmental issues in some of them. They illustrate:

- unforeseen problems despite extensive investigation;
- buried foundations missed in site investigation;
- raft foundation spanning high-wall of old quarry;
- area of poor ground not defined in site investigation;
- lack of supervision and quality control;
- mine waste and abandoned shallow mine workings;
- contamination of potable water supply.

Case history 1
Unforeseen problems despite extensive investigation

Residential development; 215 low-rise traditional masonry properties; north-west England.

The site has a well documented past. Industrial use, including a municipal waste incinerator and a civil defence depot, can be dated back to 1907 and the site is known to have been used for some waste landfilling. A limited thickness of fill, including waste material, overlies boulder clay and glacial sands and gravels. The underlying bedrock is sandstone, which the Environment Agency classes as a major aquifer. A desk study, walk-over and targeted site investigation were carried out prior to construction and identified some contamination and toxic gases. In addition to inert capping layers, gas venting and monitoring were adopted as preventative measures to reduce risks to site personnel and home-owners. During construction, significantly more contamination was found which had the potential to contaminate controlled waters and, possibly, cause significant harm to human health and the built environment. Further investigation was necessary to satisfy the Environment Agency that the controlled waters were not at risk and much more gas monitoring was subsequently carried out. In addition, continuous supervision of the construction of the capping layers was adopted to monitor the level of contamination present and increase the thickness where required.

Problem Much more contamination present than predicted from site investigation.

Impact Additional site investigation, monitoring and increased thickness of inert capping layers in discrete areas required, causing delay and additional cost.

Deficiency The amount of contamination present was underestimated. This is extremely difficult to judge and it is unlikely that more site investigation would have revealed the localised pockets of heavy contamination. However, the site history does suggest the possibility of heavy contamination; if a risk strategy had been adopted, the likelihood of these hotspots might have been predicted.

Case history 2
Buried foundations missed in site investigation

Residential development of 100 low-rise traditional masonry houses; south-west England.

The desk study and ground investigation revealed that the site had previously been used for the manufacture of sewing machines and that there had been many buildings on the site. Fill of varying thickness (0.25 m to 4 m) overlies alluvial deposits with rock head at 5 m depth. Contamination testing was carried out and the site was given a clean bill of health. A combination of foundations was chosen for the site including shallow strips, rafts and driven piles. At three of the plots where piles were planned, piling could not be carried out owing to old buried foundations forming obstructions at shallow depth in the ground. Inspection of old maps showed outbuildings had previously existed at these three locations.

Problem Old buried foundations preventing piling on poor ground.
Impact Delay and additional cost in redesigning foundations.
Deficiency Failure to report important information during desk study.

Case history 3
Raft foundation spanning high-wall of old quarry

House built on side of steep hill over old backfilled quarry; south-west England.

A local consulting engineer was employed by the builder to investigate the ground conditions and make recommendations for suitable foundations. A rudimentary site investigation revealed that the quarry had been backfilled with a variety of material including silty clayey gravel with boulders and, allegedly, domestic refuse. The depth of the fill varied considerably to a depth of 30 m. The plot was largely on the fill in the quarry, but spanned across the highwall on to the competent rock of the adjacent ground. The engineer recommended that the portion of the property over the quarry be supported on a piled raft. Immediately the property was completed, ground movements started and cracks appeared in the building. Monitoring revealed that the house and adjacent road were suffering from creep movements that may have been due to the very steeply sloping rock. It was likely that the original piles had stopped short of the bottom of the quarry and may also have been subjected to lateral forces from movement of the fill, possibly associated with collapse compression. The piles were subsequently extended to the base of the quarry but some movement has continued.

Problem Differential movement and subsidence of property.
Impact Damage to the property requiring remedial underpinning and ongoing monitoring.
Deficiency The extent of the difficulties was not fully appreciated. Expert opinion should have been sought at an early stage.

Case history 4
Area of poor ground not defined in site investigation

Medium-sized development of 23 traditional low-rise properties; north-east England.

The site had previously been occupied by an old hospital and was known to have been a quarry at an earlier stage. A local consulting engineer was employed to establish the ground conditions and make recommendations for foundations. On the basis of historical maps, the boundary of the quarry was mapped out but no ground investigation was carried out. All the plots within the quarry boundary were to be piled, all those outside were to be on shallow strip footings. During construction the boundary of the quarry was found to encroach further across the site and five more properties had to be piled.

Problem Additional properties were found to be on poor ground.
Impact Five additional properties required piling to prevent the risk of subsidence, which caused additional cost and delay.
Deficiency A lack of appreciation of the problems resulted in a failure to carry out a ground investigation. To minimise risks it is essential to research and investigate the site as much as is reasonably practicable. The additional costs of piling these five properties was far in excess of the cost of an adequate site investigation.

Case history 5
Lack of supervision and quality control

Large development of low-rise traditional masonry properties; south Wales.

The site had previously been a fossil fuel power station. Much of the site was on fill (PFA) with a maximum depth of 5 m. An extensive desk study and ground investigation, which included chemical testing and gas monitoring, revealed a minor risk to human health and the built environment. An approach was adopted based on dig and dump coupled with 1 m thick cover layers. Routine monitoring and validation testing revealed that in many cases the cover layers were not of the required thickness. All the plots were subsequently tested to verify the capping thickness and five plots had to have additional material placed.

Problem Inert cover material not of required thickness.

Impact All properties in development were checked and five properties required additional material to be placed. Additional cost of investigation and repair and delay.

Deficiency Inadequate supervision and workmanship, and failure to implement quality control or verification procedures.

Case history 6
Mine waste and abandoned shallow mine workings

Small residential development; north Wales.

The development was located in a shallow valley that was known to be backfilled by mine waste from shallow workings in the area. An intrusive investigation of five boreholes and several trial pits had been carried out on behalf of the small local developer to determine the ground conditions. Some data were collected on the ground properties. Foundation recommendations were made by the ground investigation contractor; these were adequate based on the information obtained from the investigation but, without a desk study or a mining report, the possibility of old mine workings beneath the site was overlooked. This meant that additional ground investigation was required to identify the underground voids and redesign of the foundations was necessary.

Problem Presence of voids from previous mine workings beneath development.

Impact Additional ground investigation, foundation redesign, delay and additional costs.

Deficiency Lack of desk study that would have identified the old mine workings.

Case history 7
Contamination of potable water supply

Single dwelling.

The dwelling was near a diesel fuel storage tank and its potable drinking water supply was contaminated when a small leak from one of the diesel fuel storage tanks contaminated the ground and leached through an MDPE supply pipe into the water supply. This event could have been foreseen; control measures, such as the use of impermeable pipework, could have prevented it.

Problem Leak from diesel storage tank penetrating through MDPE water pipe and into drinking water supply.

Impact Contamination of potable water supply. Extensive remedial works to water pipe system and remediation of contaminated ground around spillage.

Deficiency Lack of foresight in failing to acknowledge the likelihood of the event and the consequences. Use of wrong material for water supply pipe. Lack of contingency measures, such as an active barrier to prevent the spread of the spillage around the diesel storage tank.

Part 5 References

BRE publications, and codes and standards published by British Standards Institution, are listed separately at the end, on page 45.

Aldrich TE, Torres C and Lilquist D (1998). Risk assessment and soil contamination. *Land Contamination and Reclamation*, vol 6, no 4, pp 207-213.

Alker S, Joy V, Roberts P and Smith N (2000). The definition of brownfield. *Journal of Environmental Planning and Management*, vol 43, no 1, pp 49-69.

Andrews W C (1944). Foundations in combustible material. *The Structural Engineer*, vol 22, no 2, February, pp 53-65. Discussion on paper, June, pp 262-269.

Association of Geotechnical and Geoenvironmental Specialists (1998a). *Code of conduct for site investigation.* AGS, Beckenham, Kent.

Association of Geotechnical and Geoenvironmental Specialists (1998b). *Guidelines for good practice in site investigation.* AGS, Beckenham, Kent.

Association of Geotechnical and Geoenvironmental Specialists (1998c). *AGS Guide: the selection of geotechnical soil laboratory testing.* AGS, Beckenham, Kent.

Association of Geotechnical and Geoenvironmental Specialists (2000). Guidelines for combined geoenvironmental and geotechnical investigations. AGS, Beckenham, Kent.

Atkinson M F (1993). *Structural foundations manual for low-rise buildings.* Spon, London.

Bedford F and Smith A J (1988). Underground heating beneath Oakthorpe, Leicestershire. *Municipal Engineer*, vol 5, August, pp 167-181.

Bennett J (2002). Are you a land-condition specialist? *Proceedings of Institution of Civil Engineers, Civil Engineering*, vol 150, no 2, May, p54.

Budhu M (2000). *Soil mechanics and foundations.* Wiley, New York.

Card GB (1995). *Protecting development from methane.* Report 149. CIRIA, London.

Charles JA and Watts KS (1996). The assessment of the collapse potential of fills and its significance for building on fill. *Proceedings of Institution of Civil Engineers, Geotechnical Engineering*, vol. 119, January, pp 15-28.

Charles JA, Skinner HD and Watts KS (1998). The specification of fills to support buildings on shallow foundations: the '95% fixation'. *Ground Engineering*, vol 31, no 1, January, pp 29-33.

Charles JA and Skinner HD (2001). The delineation of building exclusion zones over hughwalls. *Ground Engineering*, vol 34, no 2, February, pp 28-33.

Charles JA and Skinner HD (2002). Compressibility of foundation fills. *Proceedings of Institution of Civil Engineers, Geotechnical Engineering*, vol. 149, no 3, pp 145-157.

Clayton CRI (2001a). Managing geotechnical risk: time for change? *Proceedings of Institution of Civil Engineers, Geotechnical Engineering*, vol 149, no 1, January, pp 3-11.

Clayton CRI (2001b). *Managing geotechnical risk – improving productivity in UK building and construction.* Thomas Telford, London. 80pp.

Clayton CRI, Matthews MC and Simons NE (1995). *Site investigation.* Blackwell Science, Oxford.

Cooling LF and Ward WH (1948). Some examples of foundation movements due to causes other than structural loads. *Proceedings of 2nd International Conference on Soil Mechanics and Foundation Engineering*, Rotterdam, vol 2, pp 162-167.

Crowhurst D and Manchester S J (1993). *The measurement of methane and associated gases from the ground.* Report 131, CIRIA, London.

Curtin WG, Shaw G, Parkinson GI and Golding JM (1994). *Structural foundation designers' manual.* Blackwell Scientific Publications, Oxford.

Department of the Environment (1996). Major coalfields regeneration plans unveiled. News release no 467, 6th November.

Department of the Environment, Transport and the Regions (2000a). *Planning policy guidance note PPG 3: housing.* The Stationery Office, London.

Department of the Environment, Transport and the Regions (2000b). *Our towns and cities: the future – delivering an urban renaissance.* Urban White Paper.

Department of the Environment, Transport and the Regions (2000c). *Guidelines for environmental risk assessment and management – revised departmental guidance.* Stationery Office, London.

Department of the Environment, Transport and the Regions (2000d). *Building Regulations 2000.* Stationery Office, London.

Environment Agency and NHBC (2000). *Guidance for the safe development of housing on land affected by contamination.* R&D Publication 66. The Stationery Office, London. 86pp.

Environmental Industries Commission in association with the Laboratory of the Government Chemist (1997). *A quality approach for contaminated land consultancy.* Contaminated Land Research Report CLR 12, Department of the Environment.

Garvin S, Hartless R, Smith MA, Manchester S and Tedd P (1999). *Risks of contaminated land to buildings, building materials and services: a literature review.* Environment Agency R&D Technical Report P331. WRc, Swindon.

Geen B (1944). Fire under a factory. *Journal of Institution of Civil Engineers*, vol 23, pp 91-93.

Greenwood DA (1986). Discussion. Building on Marginal and Derelict Land. *Proceedings of Institution of Civil Engineers conference*, Glasgow, pp 397-398. Thomas Telford, London.

Harries CR, Witherington PJ and McEntee JM (1995). *Interpreting measurements of gas in the ground.* Report 151, CIRIA, London.

Harris MR, Herbert SM and Smith MA (1995). *Remedial treatment for contaminated land, vol 6, Containment and hydraulic measures.* Special Publication 106, CIRIA, London.

Hatem D J [ed] (1998). Subsurface conditions – risk management for design and construction management professionals. Wiley, New York.

Health and Safety Executive (1985). *The Abbeystead explosion – a report of the investigation by the Health and Safety Executive into the explosion on 23 May 1984 at the valve house of the Lune/Wyre water transfer scheme at Abbeystead.* HMSO, London.

Health and Safety Executive (1998). *Managing health and safety – five steps to success.* HSE Books, London.

Health and Safety Executive (1999). *Five steps to risk assessment.* HSE Books, London.

Health and Safety Executive (2001). *Reducing risks, protecting people.* (Discussion document published in 1999). HSE Books, Sudbury, Suffolk.

Healy PR and Head JM (1984). *Construction over abandoned mine workings.* Special Publication 32, CIRIA , London.

Interdepartmental Committee on the Redevelopment of Contaminated Land (1986). *Notes on the fire hazard of contaminated land.* Note 61/84, 2nd edition.

Institution of Civil Engineers (1999). *Specification for cement-bentonite cut-off walls.* Thomas Telford, London.

Institution of Civil Engineers Site Investigation Steering Group (1993). *Site investigation in construction.* Thomas Telford, London.

Jarvis ST and Brooks TG (1996). The use of pfa/cement pastes in the stabilisation of abandoned mine workings. *Waste Management*, vol 16, pp 135-143.

Lambe TW and Whitman RV (1979). *Soil mechanics*, SI version. Wiley, New York.

Lampert D and Woodley DR (eds) (1991). *Site selection and investigation: a practical handbook.* Gower, Aldershot.

Law Society (1998). *Coal mining searches - England and Wales: guidance notes and directory.*

Law Society of Scotland (1999). *Coal mining searches - guidance notes and directory.*

Martin WS (1996). *Site guide to foundation construction. - a handbook for young professionals.* CIRIA Special Publication 136.

McDowell PW, Barker RD, Butcher AP, Culshaw MG, Jackson PD, McCann DM, Skipp BO, Matthews SL and Arthur JCR (2002). *Geophysics in engineering investigations.* Report C562, CIRIA, London.

National House Building Council (1999). *NHBC Standards Chapter 4.1: Land quality – managing ground conditions.*

Nottingham Trent University (1994). *Sampling strategies for contaminated land.* Contaminated Land Research Report CLR 4, Department of the Environment.

O'Riordan NJ and Milloy CJ (1995). *Risk assessment for methane and other gases from the ground.* CIRIA Report 152.

Palmer KN (1979). Fire and explosion hazards with contaminated land. Reclamation of contaminated land. *Proceedings of conference, Eastbourne*, October 1979, pp D3/1-7. Society of Chemical Industry, London, 1980.

Parliamentary Office of Science and Technology (1998). *A brown and pleasant land – household growth and brownfield sites.* Report 117, July. House of Commons, London.

Privett KD, Matthews SC and Hodges RA (1996). *Barriers, liners and cover systems for containment and control of land contamination.* Special Publication 124, CIRIA, London.

Raybould JG, Rowan SP and Barry DL (1995). *Methane investigation strategies.* Report 150, CIRIA, London.

Royal Society (1992). R*isk: analysis, perception and management.* Report of a Royal Society Study Group.

Sear LKA (2001). *The properties and use of coal fly ash.* Thomas Telford, London.

Seco e Pinto PS [ed] (1998). Environmental geotechnics. *Proceedings of 3rd Internat-ional Congress, Lisbon, 4 vols. Balkema, Rotterdam.*

Skinner HD, Watts KS and Charles JA (1997). Building on colliery spoil: some geotechnical considerations. *Ground Engineering*, vol. 30, no 5, June, pp 35-40.

Sleep K (1996). Coal deal will yield £1 billion in work. *Construction News*, 14th November, p6.

SNIFFER (1999). *Communicating understanding of contaminated land risks.* Final report. Scotland and Northern Ireland Forum for Environmental Research.

Stationery Office (1999). *The Water Supply (Water Fittings) Regulations 1999.* Statutory Instrument 1999 No 1148.

Stephens J and Norris M (1994). *Laying potable water pipelines in contaminated ground. - guidance note.* FWR report FR 0448. Water Research Centre.

Summersgill M (1997). Management of risks in geoenvironmental investigations. Geoenvironmental engineering - contaminated ground: fate of pollutants and remediation (eds R N Yong and H R Thomas). *Proceedings of British Geotechnical Society conference*, Cardiff, pp 544-550. Thomas Telford, London.

Taunton P and Adams R (2001). Impact management during remediation of a former landfill in an urban environment. Geoenvironmental engineering – geoenvironmental impact management (eds R N Yong and H R Thomas). *Proceedings of 3rd British Geotechnical Association conference*, Edinburgh, pp 251-256. Thomas Telford, London.

Tedd P (2001). Field investigations of slurry trench cut-off walls to control pollution migration. *Proceedings of 3rd British Geotechnical Association conference*, Edinburgh, pp 575-582. Thomas Telford, London.

Tedd P, Holton IR, Butcher AP, Wallace S and Daly PJ (1997). Investigation of the performance of cement-bentonite cut-off wall in aggressive ground at a disused gasworks site. *Proceedings of International Containment Technology Conference*, St Petersburg, Florida, pp 125-132. Also Land Containment & Reclamation, vol 5, no 3, pp 217-223.

Thompson RP (1998). The value of timely hazard identification. The value of geotechnics in construction. *Proceedings of seminar*, London, pp 3-11. Construction Research Communications.

Tomlinson MJ (1995). *Foundation design and construction* (6th edition). Longman Scientific and Technical, Harlow, Essex.

Trenter NA and Charles JA (1996). A model specification for engineered fills for building purposes. *Proceedings of Institution of Civil Engineers*, vol 119, no 4, October, pp 219-230.

Trew JE, Tarbet NK, De Rosa PJ, Morris D, Cant J and Olliff JL (1995). *Pipe materials selection manual – water supply* (2nd edition). Water Research Centre.

Uff JF and Clayton CR I(1986). *Recommendations for the procurement of ground investigation.* Special Publication 45, CIRIA, London.

Urban Task Force (1999). *Towards an urban renaissance.* Spon, Andover, Hampshire.

Vadgama NJ (1986). Discussion. Building on Marginal and Derelict Land. *Proceedings of Institution of Civil Engineers conference*, Glasgow, pp 399-401. Thomas Telford, London.

Watts KS and Charles JA (1997). Treatment of compressible foundation soils. Ground improvement geosystems: densification and reinforcement. *Proceedings of 3rd international conference*, London, June, pp 109-119.

Watts KS and Charles JA (1999). Settlement characteristics of landfill wastes. *Proceedings of Institution of Civil Engineers*, Geotechnical Engineering, vol. 137, October, pp 225-233.

Weatherley N (1979). Trench filled PFA in colliery waste supporting old people's bungalows. Engineering behaviour of industrial and urban fill. Proceedings of Midland Geotechnical Society Symposium, pp D71-D76.

Westcott FJ, Lean CMB and Cunningham ML (2001a). *Piling and penetrative ground improvement methods on land affected by contamination: interim guidance on pollution prevention.* Environment Agency, National Groundwater and Contaminated Land Centre Project NC/99/73.

Westcott FJ, Smith JWN and Lean CMB (2001b). Piling on land affected by contamination: environmental impacts, regulatory concerns and effective solutions. Geoenvironmental engineering – geoenvironmental impact management. *Proceedings of 3rd British Geotechnical Association conference*, Edinburgh, pp 103-108. Thomas Telford, London.

Williams GM and Aitkenhead N (1991). Lessons from Loscoe: the uncontrolled migration of landfill gas. *Quarterly Journal of Engineering Geology*, vol 24, pp 191-207.

Wood AA and Griffiths CM (1994). Debate: contaminated sites are being over-engineered. *Proceedings of Institution of Civil Engineers, Civil Engineering*, vol. 102, no 3, August, pp 97-105.

Yong RN and Thomas HR [eds] (1997). Geoenvironmental engineering - contaminated ground: fate of pollutants and remediation. *Proceedings of British Geotechnical Society conference*, Cardiff. Thomas Telford, London.

Yong RN and Thomas HR [eds] (2001). Geoenvironmental engineering – geoenviron-mental impact management. *Proceedings of 3rd British Geotechnical Association conference*, Edinburgh. Thomas Telford, London.

BRE Digests

276 Hardcore
318 Site investigation for low-rise building: desk studies
322 Site investigation for low-rise building: procurement
348 Site investigation for low-rise building: the walk-over survey
363 Sulfate and acid resistance of concrete in the ground
Revised as Special Digest 1 - see below
381 Site investigation for low-rise building: trial pits
383 Site investigation for low-rise building: soil description
395 Slurry trench cut-off walls to contain contamination
411 Site investigation for low-rise building: direct investigations.
427 Low-rise buildings on fill
Part 1 Classification and load-carrying characteristics
Part 2 Site investigation, ground movement and foundation design
Part 3 Engineered fill
Special Digest 1 Concrete in aggressive ground
Part 1 Assessing the chemical environment
Part 2 Specifying concrete and additional protective measures
Part 3 Design guide for common applications
Part 4 Design guide for precast products

BRE Information Papers

2/87 Fire and explosion hazards associated with the redevelopment of contaminated land
3/89 Subterranean fires in the UK – the problem

BRE Reports

BR 100 Measurement of gas emissions from contaminated land
BR 211 Radon: guidance on protective measures for new dwellings
BR 212 Construction of new buildings on gas-contaminated land
BR 255 The performance of building materials in contaminated land
BR 391 Specifying vibro stone columns
BR 414 Protective measures for housing on gas-contaminated land
BR 424 Building on fill: geotechnical aspects

British Standards Institution

BS 1377:1990 Methods of test for soils for civil engineering purposes: Parts 1-9
BS 1047:1983 Specification for air-cooled blastfurnace slag aggregate for use in construction
BS 5930:1999 Code of practice for site investigations
BS 8004:1986 Code of practice for foundations
BS 10175:2001 Investigation of potentially contaminated sites
DD ENV 1997 Eurocode 7. Geotechnical design
DD ENV 1997-1: 1995 General rules
DD ENV 1997-2: 2000 Design assisted by laboratory testing
DD ENV 1997-3: 2000 Design assisted by field testing